T0135364

Springer Theses

Recognizing Outstanding Ph.D. Research

Aims and Scope

The series "Springer Theses" brings together a selection of the very best Ph.D. theses from around the world and across the physical sciences. Nominated and endorsed by two recognized specialists, each published volume has been selected for its scientific excellence and the high impact of its contents for the pertinent field of research. For greater accessibility to non-specialists, the published versions include an extended introduction, as well as a foreword by the student's supervisor explaining the special relevance of the work for the field. As a whole, the series will provide a valuable resource both for newcomers to the research fields described, and for other scientists seeking detailed background information on special questions. Finally, it provides an accredited documentation of the valuable contributions made by today's younger generation of scientists.

Theses are accepted into the series by invited nomination only and must fulfill all of the following criteria

- They must be written in good English.
- The topic should fall within the confines of Chemistry, Physics, Earth Sciences, Engineering and related interdisciplinary fields such as Materials, Nanoscience, Chemical Engineering, Complex Systems and Biophysics.
- The work reported in the thesis must represent a significant scientific advance.
- If the thesis includes previously published material, permission to reproduce this must be gained from the respective copyright holder.
- They must have been examined and passed during the 12 months prior to nomination.
- Each thesis should include a foreword by the supervisor outlining the significance of its content.
- The theses should have a clearly defined structure including an introduction accessible to scientists not expert in that particular field.

More information about this series at http://www.springer.com/series/8790

Thuy T. Pham

Applying Machine Learning for Automated Classification of Biomedical Data in Subject-Independent Settings

Doctoral Thesis accepted by
The University of Sydney, Australia

 Springer

Author
Dr. Thuy T. Pham
School of Electrical and Information
    Engineering
The University of Sydney
Sydney, NSW, Australia

Supervisor
Prof. Philip H. W. Leong
School of Electrical and Information
    Engineering
The University of Sydney
Sydney, NSW, Australia

ISSN 2190-5053   ISSN 2190-5061   (electronic)
Springer Theses
ISBN 978-3-030-07518-7   ISBN 978-3-319-98675-3   (eBook)
https://doi.org/10.1007/978-3-319-98675-3

This Springer imprint is published by the registered company Springer Nature Switzerland AG
The registered company address is: Gewerbestrasse 11, 6330 Cham, Switzerland

To my beloved brother, Tien Van Pham, you would have been an author of a thesis if you could bounce back from your sickness. My Ph.D. should have been yours but I wish I made up for opportunities you missed.

Supervisor's Foreword

Research in machine learning has reached a point where engineered systems are capable of surpassing human performance on complex cognitive tasks. This has been recently demonstrated in a number of challenging domains including self-driving cars, automatic language translation and strategy games. Machine learning techniques promise to revolutionise all areas of technology and provide significant societal benefits.

One of the most challenging problems is health care as existing processes, equipment and techniques are struggling to scale with increasing demands. Machine learning has significant promise, and it was with this in mind that Dr. Pham and I set the ambitious goal of developing general techniques to improve machine learning accuracy in biomedical signal processing applications.

Machine learning is concerned with extracting models from data. Identifying salient features is key to achieving good performance. The key contribution of this work is a methodology to identify features which are more discriminative and better correlated with the target than before. This was achieved through the generation of a superset of candidate features and then reducing its size through a selection process. To demonstrate its effectiveness, domain-specific ways to utilise the resulting features to address problems in human movement assessment, respiratory artefact removal and spike sorting of electrophysiological data were derived. Remarkably, Dr. Pham was able to achieve state-of-the-art performance for all of these disparate applications.

Working with Dr. Pham was an amazing experience which I enjoyed very much. I hope that you also will enjoy this book, and that it will inspire new ideas that improve society.

Sydney, Australia
March 2018

Philip H. W. Leong
Professor of Computer Systems

Parts of this thesis have been published in the following articles:

Journals

- Pham TT, Leong PH, Robinson PD, Gutzler T, Jee AS, King GG, Thamrin C (2017), "Automated quality control of forced oscillation measurements: respiratory artifact detection with advanced feature extraction," Journal of Applied Physiology (JAP) 123(4):781789.
- Pham TT, Moore ST, Lewis SJG, Nguyen DN, Dutkiewicz E, Fuglevand AJ, McEwan AL, Leong PH (2017), "Freezing of gait detection in Parkinson's disease: a subject-independent detector using anomaly scores," IEEE Transactions on Biomedical Engineering (IEEE TBME) 64(11):27192728.
- Pham TT, Thamrin C, Robinson PD, McEwan A, Leong PH (2016) "Respiratory artefact removal in forced oscillation measurements: A machine learning approach," IEEE Transactions on Biomedical Engineering (IEEE TBME) 64 (7):19. https://doi.org/10.1109/TBME.2016.2554599.

Conferences

- Thuy T. Pham, Diep N. Nguyen, Eryk Dutkiewicz, Alistair L. McEwan, and Philip H.W. Leong, "Wearable Healthcare Systems: A Single Channel Accelerometer Based Anomaly Detector for Studies of Gait Freezing in Parkinson's Disease," ICC'17 SAC-6 EH, Paris, France, May 2017.
- Thuy T. Pham, Diep N. Nguyen, Eryk Dutkiewicz, Alistair L. McEwan, Cindy Thamrin, Paul D. Robinson, Philip H.W. Leong, "Feature Engineering and Supervised Learning Classifiers for Respiratory Artefact Removal in Lung Function Tests," *IEEE Globecom-SAC-EH 2016*, Washington DC, USA Dec 2016.
- Thuy T. Pham, Andrew J. Fuglevand, Alistair L. McEwan, and Philip H. W. Leong, "Unsupervised Discrimination of Motor Unit Action Potentials Using Spectrograms," Engineering in Medicine and Biology Society (EMBC), 36th Annual International IEEE EMBS Conference, Chicago USA. Conf Proc IEEE Eng Med Biol Soc 2014;2014:1-4.

Acknowledgements

This thesis presents research results carried out at the Computer Engineering Laboratory (CEL, The University of Sydney, Australia) with the collaboration of Woolcock Medical Institute (Glebe, NSW, Australia), Westmead Children Hospital (NSW, Australia), Brain and Mind Centre (University of Sydney, NSW, Australia) and Icahn School of Medicine at Mount Sinai, NY, USA.

I would like to express my thankfulness to my advisors Prof. Philip H. W. Leong and Associate Prof. Alistair L. McEwan for their advice and encouragement. This thesis would have been impossible without their guidance.

I am indebted to Dr. Cindy Thamrin (Senior R.F. at Woolcock, University of Sydney) and Dr. Paul Robinson (Senior Lecturer at University of Sydney, Westmead Hospital) who have provided medical insights for lung function test data projects.

I would like to thank Prof. Simon Lewis (Medical School, University of Sydney) and Prof. Steven Moore (Icahn Medical School, NY, USA) who have supported my gaiting freezing project for patients with Parkinson's disease.

I also highly appreciate Prof. Dutkiewicz, Dr. Diep Nguyen, Australian government (Endeavour/Prime Minister Award) and the Faculty Research Cluster Program at The University of Sydney for financial supports and advising during my study.

I gratefully acknowledge the help from Dr. Noorian for proofreading this thesis and other CEL lab members for their valuable discussions in machine learning.

I would love say thanks to my parents (Toan Van Pham, Them Thi Nguyen), my brother (Tien Van Pham), my husband and my children (Veronica and Paul) who gave me the strength and self-confidence to pursue my goals of life.

Contents

1 Introduction 1
  1.1 Motivation and Approach 1
  1.2 Contributions 2
  1.3 Thesis Structure 3
  References 3

2 Background 5
  2.1 Unsupervised Classification 5
    2.1.1 Multi-class Classification 5
    2.1.2 Two-Class Sorting: Anomaly Detection 7
  2.2 Performance Metrics 7
  2.3 Feature Engineering 10
    2.3.1 Feature Relevance 11
    2.3.2 Feature Selection 11
    2.3.3 Automatic Selection Models 12
  2.4 Summary 15
  References 15

3 Algorithms 19
  3.1 Feature Engineering 19
    3.1.1 Automated Selection Process 19
    3.1.2 Saliency Criteria 20
  3.2 Classification Algorithms 22
    3.2.1 Point Anomaly Detection Scheme 22
    3.2.2 Collective Anomaly Detection Scheme 22
    3.2.3 Correlation Based Spike Sorting Scheme 23
  3.3 Summary 24
  References 25

4 Point Anomaly Detection: Application to Freezing of Gait Monitoring 27
4.1 Background on Gait Freezing Detection in Parkinson's Disease 27
4.2 FoG Detector 29
4.3 Data Set 30
4.3.1 Development Set 30
4.3.2 Test Set 32
4.4 Feature Extraction 33
4.4.1 Existing Features 33
4.4.2 New Features 34
4.4.3 Anomaly Scores 35
4.4.4 Exploratory Pool 35
4.4.5 Feature Selection 37
4.5 Performance Metrics 38
4.6 Results 39
4.6.1 Selection by Saliency 39
4.6.2 Selection By Robustness 39
4.6.3 Selection By Detection Performance 41
4.6.4 Tests and Comparisons with the Same Cohort Set 43
4.6.5 External Validation Tests 44
4.7 Discussion 45
4.8 Summary 45
References 46

5 Collective Anomaly Detection: Application to Respiratory Artefact Removals 49
5.1 Background on Respiratory Artefact Removal in FOT Data 49
5.2 Data Collection 50
5.2.1 Subjects and Protocol 50
5.2.2 Data Pre-processing 51
5.3 Performance Metrics 52
5.4 Proposed Artefact Detection Scheme 53
5.5 Feature Extraction 53
5.5.1 Feature Pool 54
5.5.2 Challenging Factors and Other Criteria 55
5.6 Unsupervised Artefact Detector 58
5.6.1 Single Filter Approach 58
5.6.2 Multi-filter Approach 58
5.7 Supervised Learning Artefact Detector 60
5.7.1 Machine Learning and Challenges 60
5.7.2 Feature Extraction 62
5.7.3 Feature Learning and Supervised Classifier 62

5.8   Results..   63
        5.8.1   Saliency Ranking...   63
        5.8.2   Unsupervised Artefact Detector: Single Filter Results....   64
        5.8.3   Unsupervised Artefact Detector: Multi-filter Results.....   69
        5.8.4   Supervised Artefact Detector: Machine Learning
                 Classifier Results..   73
5.9   Discussion...   76
5.10   Summary...   78
References...   79

6 **Spike Sorting: Application to Motor Unit Action Potential**
   Discrimination...   83
        6.1   Background on Electromyography Motor Unit Analysis.......   83
                6.1.1   MUAPs......................................   83
                6.1.2   Spike Sorting..............................   83
        6.2   Data Collection...   84
                6.2.1   Physiologically Based Synthetic Data...............   84
                6.2.2   Recorded Data..............................   84
        6.3   Feature Pool..   85
        6.4   Automated Spike Sorter.....................................   86
                6.4.1   Preprocessing.............................   86
                6.4.2   x-Class Sorter.............................   87
        6.5   Reference Works...   87
        6.6   Performance Metrics...   88
        6.7   Results...   88
                6.7.1   Selected Features..........................   88
                6.7.2   Sorting Performance........................   89
        6.8   Summary...   93
        References...   94

7 **Conclusion**...   97
        7.1   Proposed Algorithms...   97
                7.1.1   Feature Engineering........................   98
                7.1.2   Classifiers................................   98
        7.2   Experiment Results..   98
                7.2.1   Point Anomaly Detection Application................   98
                7.2.2   Collective Anomaly Detection Application...........   99
                7.2.3   Spike Sorting Application..................   100
        7.3   Summary...   100
        References...   101

Appendix A: Wavelet Decomposition and Spectral Coherence........   103

Appendix B: Table of Settings for Synthetic nEMG Data
             in Chapter 6...................................   105

About the Author...   107

Chapter 1
Introduction

1.1 Motivation and Approach

This thesis argues that the classification performance of unsupervised and subject-independent automated sorters for biomedical data can be improved by exploiting data-driven and domain-knowledge-driven strategies that help find better features and more efficient sorters.

In the first scenario, accelerometry data were used to assess body movements, specifically to make a binary classification for freezing of gait (FoG) or normal events over a number of FoG is one of the most common symptoms of Parkinson's disease (PD) and strongly relates to falls. Objective FoG detection has been a pressing concern, particularly out-of-lab deployment with wearable devices. Current automated methods have been proposed with various *global* parameters (i.e., inconsistent threshold fixed values and/or different data channel settings found in literature). This suggests a high variability in actual thresholds over time and subjects.

The second scenario, which is also a two-class discrimination problem, involves removing respiratory artefacts in the forced oscillation technique (FOT). The averages of measurements in lung function tests (e.g., total respiratory mechanical resistance) are the main outcomes in clinical and research usage, which are significantly affected by the artefacts. Consequently, more work is required to improve the reproducibility of FOT by automatically eliminating respiratory artefacts. Apart from the natural dependency of breath samples on time and subjects, we found that the normality of given data should not be assumed as it has been rejected by common test statistics Hence, besides choosing better features, more general statistical parameters with quartiles should be applied rather than existing methods with the normality assumption.

Thirdly, multi-class sorting for intramuscular electromyography (nEMG) *spikes* (action potentials) can help identify classes that are often referred to motor units (MUs). Single motor unit activity study is a major research interest because changes of single motor unit activities (e.g., MU action potential (MUAP) morphology, MU activation, and MU recruitment) provide the most informative part in diagnosis and treatment of neuromuscular disorders. Nevertheless, nEMG data often provide more than one MU activities, thus MUAP discrimination is a crucial task to study single

© Springer Nature Switzerland AG 2019
T. T. Pham, *Applying Machine Learning for Automated Classification of Biomedical Data in Subject-Independent Settings*, Springer Theses, https://doi.org/10.1007/978-3-319-98675-3_1

unit activities. One important note is that the number of classes in this classification task could not be pre-defined. Hence clustering methods are often employed. Existing features have been calculated from Euclidean distances which assumes a spherical distribution of data. To account for electrode drift and normalized values that suit subject-independent settings, we proposed to use the correlation metric that range $-1 \rightarrow 1$.

These three cases share certain difficulties. One is that the de facto standard practice of each relies on human-based assessment which is almost always subjective and time-consuming. Another is that it is challenging to assess the relevance and clusterability of existing features. Higher correlated and more separable features across classes may improve the classification performance of subject-independent classifiers. We have reviewed the state of the art in each above case and found that most existing automated efforts have failed to address the aforementioned factors. Those approaches often have been designed primarily in subject-dependent settings to yield excellent accuracy performance. Meanwhile we aim to launch unsupervised methods for subject-independent settings. We use unsupervised classifiers in the process of finding more salient and discriminative features. Once these applied machine learning methods are used, unsupervised manners would alternate the laborious subjective manual methods.

1.2 Contributions

This thesis aims to improve accuracy performance of classification tasks in subject-independent settings by utilsing supervised techniques to find better features (i.e., more discriminative and higher correlated with the desired output). A voting-based technique has been proposed to analyze ranking scores by several saliency criteria including mutual information, Euclidean distance based discrimination, and variance ratio based clusterability. This hybrid selection scheme is a data-driven approach and can compare a comprehensive set of candidates including existing features and novel variants. Given a large set of exploratory feature candidates, the most selective features learnt from this process are most applicable to the unsupervised and subject-independent applications. Exploiting this strategy in each scenario, better models are also suggested through this domain-knowledge-driven approach (e.g., issues associated with dependency in Case 2 and/or other related domain knowledge in Cases 1 and 3). This approach is applicable to a wide range of machine learning applications as well.

The main contributions of this work are:

1. This is the first reported feature selection technique based on voting which considers not only mutual information criterion but also clusterability for respiratory artefact removal in FOT measurements [2–4], FoG detection [5, 7], and nEMG spike sorting [1].

2. Novel features have been found through this work are more relevant and discriminative in the FoG, FOT, and nEMG data than the existing ones [1, 3, 5].
3. Propose anomaly detectors which, to the best of our knowledge, achieve the best reported performance for unsupervised subject-independent settings for FOT data regardless of participants' age [4] and FoG data [6].
4. Suggest an efficient unsupervised spike sorting when the class number is not pre-defined for subject-independent settings [1].

1.3 Thesis Structure

The thesis is organized in seven chapters. Chap. 2 provides a background and literature review for the three applications: FoG detection with proper acceleration data, respiratory artefact removal in FOT data, and MUAP sorting for nEMG data. In Chap. 3, algorithms including feature engineering and sorting schemes are described. Details of data collection, parameter setting, and experiment results are demonstrated separately for each specific domain such as FoG detection (Chap. 4, Point anomalies), FOT respiratory artefact removal (Chap. 5, collective anomalies), and spike sorting for nEMG data (Chap. 6). The final chapter concludes with a summary and future work for this thesis.

References

1. Pham TT, Fuglevand AJ, McEwan AL, Leong PH (2014) Unsupervised discrimination of motor unit action potentials using spectrograms. In: 36th Annual international conference of the IEEE engineering in medicine and biology society (EMBC) 2014, IEEE, pp 1–4
2. Pham TT, Nguyen DN, Dutkiewicz E, McEwan AL, Thamrin C, Robinson PD, Leong PH (2016a) Feature engineering and supervised learning classifiers for respiratory artefact removal in lung function tests. In: Global communications conference (GLOBECOM), 2016 IEEE, IEEE, pp 1–6
3. Pham TT, Thamrin C, Robinson PD, McEwan A, Leong PH (2016b) Respiratory artefact removal in forced oscillation measurements: A machine learning approach. IEEE Trans Biomed Eng 64(7):1–9
4. Pham TT, Leong PH, Robinson PD, Gutzler T, Jee AS, King GG, Thamrin C (2017a) Automated quality control of forced oscillation measurements: respiratory artifact detection with advanced feature extraction. J Appl Physiol 123(4):781–789
5. Pham TT, Moore ST, Lewis SJG, Nguyen DN, Dutkiewicz E, Fuglevand AJ, McEwan AL, Leong PH (2017b) Freezing of gait detection in Parkinson's disease: a subject-independent detector using anomaly scores. IEEE Trans Biomed Eng 64(11):2719–2728
6. Pham TT, Nguyen DN, Dutkiewicz E, McEwan AL, Leong PH (2017c) Wearable healthcare systems: a single channel accelerometer based anomaly detector for studies of gait freezing in Parkinson's disease. In: 2017 IEEE International conference on communications (ICC), IEEE, pp 1–5

Chapter 2
Background

2.1 Unsupervised Classification

A classification task involves finding a mapping from features to a categorical variable. When no label in the training phase is used, the task is referred to as unsupervised classification. There are two common problems for this: anomaly detection (i.e., two-class) and clustering or x-class sorting with an unknown number of classes.

Let input data $D = x_1, \ldots, x_N$ where N is number of data points. D is separated into k disjoint subsets $C_1, \ldots, C_k$ ($k \ll N$): $C_i \cap C_j = \emptyset$ if $i \neq j$ and $D = C_1 \cup \ldots \cup C_k$. The result of clustering depends on a measure of similarity between the elements and the aim is to place similar elements in the same cluster.

2.1.1 Multi-class Classification

The basic idea of multi-class sorting is to group *similar* instances, based on some *distance* metric. Several ways to establish the similarity between data points are commonly used including: Euclidean distances among the group members, dense areas of the data space, or high correlation coefficients. For biomedical data, an example is grouping needle EMG motor unit action potentials (MUAP) in order to find the number of active motor units, i.e., discover the number of clusters, k, in spike sorting [40].

In the spike sorting literature, algorithms using distances (e.g., k-means clustering [30, 50], mean shift [11, 56]), likelihood (e.g., Bayesian classification (BC) [10]) and others: template matching [8], neural network based [37], super paramagnetic clustering (SPC) [4], or density grid contour clustering [54] have all been proposed.

The k-means algorithm is a classic model for clustering, originally developed for vector quantization [30]. The k-means algorithm requires k to be given a priori from a set of M points $x_1, \ldots, x_M$ in order to find a label variable $y_1, \ldots, y_M$

© Springer Nature Switzerland AG 2019
T. T. Pham, *Applying Machine Learning for Automated Classification of Biomedical Data in Subject-Independent Settings*, Springer Theses, https://doi.org/10.1007/978-3-319-98675-3_2

where $y_i \in 1, \ldots, k$. First, it randomly initializes centroids $\mu_1, \mu_2, \ldots, \mu_k$. It then calculates the *least-squares* cost of this initial arrangement using k mean vectors $c_1, \ldots, c_k$. Then it assigns a cluster label y_i to each point x_i and recomputes the centroids until the following optimization criterion is met.

$$y_1, \ldots, y_m = \underset{c_1,\ldots,c_k}{\mathrm{argmin}} \sum_{j=1}^{k} \sum_{y_i=j} \|x_i - c_j\|^2 \qquad (2.1)$$

Another popular multi-class algorithm is *super paramagnetic clustering* (SPC) [4], a state-of-the-art method, has been used to launch a spike-sorting module for neural data (e.g., [4, 39]). The SPC method uses interactions between a data point (a spike) and its k-nearest neighbours [4]. If the interactions are strong, spikes are more similar. Refinement is implemented as a Monte Carlo iteration of a Potts model [44] which suggests the behaviour of ferromagnets and certain other phenomena of solid-state physics.

The term *temperature* in SPC is used to interpret the probability at which the states of a number of neighbouring data points change simultaneously [46]. At a relatively high temperature, all the points switch randomly, regardless of their interactions (paramagnetic phase). At a low temperature, all the points change their states together (ferromagnetic phase). At medium temperatures (super paramagnetic phase) only points in the same group change their states concurrently. In a clustering application, the ferromagnetic phase, the paramagnetic phase, and the super paramagnetic phase can be considered a classifying result of one single cluster, several tiny clusters, and a number of medium-size clusters, respectively.

First, SPC represents m features of a spike i by a point x_i in an m-dimensional space. Then it finds the interaction strengths between the point x_i and k nearest neighbouring points. The interaction strength J_{ij} between x_i and one of its neighbours, named x_j, is given by [46]. From Eq. (2.2), J_{ij} reduces exponentially when the Euclidean distance $\|x_i - x_j\|^2$ increases. A smaller distance results in a stronger similarity between two spikes.

$$J_{ij} = \begin{cases} \frac{1}{k} \exp(-\frac{\|x_i - x_j\|^2}{2a^2}) & \text{if } x_i \text{ is one of k nearest neighbors of } x_j \\ 0 & \text{otherwise.} \end{cases} \qquad (2.2)$$

where a is the average distance from x_i to its k nearest neighbours.

Then, SPC assigns each point x_i to a random state s in a set of q states. N Monte Carlo iterations are run for different temperatures using the Swendnsen-Wang algorithm [5] or the Wolff algorithm [55]. Blatt et al. [5] recommended a setting of $q = 20$ states, $k = 11$ nearest neighbours, and $N = 500$ iterations for clustering. With this setting, the clustering process would mainly depend on the temperature parameter and is robust to small changes of other parameters.

Though there have been enormous number of cluster analysis algorithm proposals across research areas, the most suitable algorithm for a particular problem is often chosen by experimentation [17]. In an early review by Lewicki [28] and Gibson et al. [18], the current benchmark method is k-means clustering [50] because it is simple and fast but requires an assumption of the given k. The valley-seeking [58] and super paramagnetic clustering (SPC) [4] are *cutting edge* methods but have high computational complexity [18]. For example, SPC uses default settings of 100 Monte Carlo iterations, increasing computation time by several order of magnitude. SPC also involves an estimation of the upper bound of k in the settings [39]. Most of these techniques use the Euclidean distance metric, that assumes a spherical distribution of data. Due to the effect of electrode drift, ellipsoidal clusters are formed in practice, not spherical [18]. In this work, an alternative approach is proposed.

2.1.2 Two-Class Sorting: Anomaly Detection

Anomaly detection is a special case of sorting when $k = 2$. $D = C_1 \cup C_2$ where C_1 is the anomaly set that are considerably dissimilar from the remainder, C_2. In other words, one class is for all *normal* data point, the other is for all *anomalies*. An anomaly is a deviation from the normal or expected behaviour.

In the literature, C_1 is referred to as outliers, exceptions, peculiarities, noise, or novelties [21, 27, 51]. If the data points are merely unwanted and not of interest to the study, this task can be considered noise removal (e.g., robust regression and outlier detection [21, 27]). For monitoring behaviours, it is often called novelty detection (e.g., unusual user behaviour and unrecognised activities [51]), or the converse, anomaly detection. In this thesis two real life examples in biomedical data are studied: freezing of gait events (Case 1) and respiratory artefact cycles (Case 2).

Key challenges found in this process include defining a representative normal set (i.e., C_2) that is hard and domain specific, and the boundary between C_1 and C_2 is not being always precise. Furthermore, in the normal set, element behaviour may be evolving. Thus, according to problem characteristics (i.e., nature of data, anomaly type, labels, and output) there are diverse ways to detect anomalies. Recently, several approaches including supervised-classification-based (e.g., rule-based, neural networks, or support vector machine based), clustering based, statistical (e.g., parametric or non-parametric), information theory, and visualization based were reviewed in a survey [9]. Similar to comments in the previous section, in order to be employed towards wearable device and real-time detection, simple thresholding algorithms using statical parameters are preferred [41–43].

2.2 Performance Metrics

In a multi-class classification task, the confusion matrix provides a summary of the performance achieved by a classifier. Let C be the number of classes. The confusion matrix M is a square matrix of $C \times C$ where $M_{i,j}$ denotes the number of test outcomes (i.e., *ground truth* labels, G_i) of class i, that were predicted as class j, P_i (Eq. (2.3)). The number of successful predicted events (*True*) for class i, denoted T_{ii}, is the diagonal line of M. All other members of M are incorrectly predicted events (*False*), denoted F_{ij} where $i \neq j$.

$$M = \begin{matrix} & P_1 & \cdots & P_i & \cdots & P_C & \\ \begin{pmatrix} T_{11} & \cdots & F_{1i} & \cdots & F_{1C} \\ \vdots & \ddots & \vdots & \cdots & \vdots \\ F_{i1} & \cdots & T_{ii} & \cdots & F_{iC} \\ \vdots & \cdots & \vdots & \ddots & \vdots \\ F_{C1} & \cdots & F_{Ci} & \cdots & T_{CC} \end{pmatrix} & \begin{matrix} G_1 \\ \vdots \\ G_i \\ \vdots \\ G_C \end{matrix} \end{matrix} \tag{2.3}$$

The sensitivity and positive predictive value (PPV) of class i, Sen_i and PPV_i, are defined by:

$$\text{Sen}_i = \frac{T_{ii}}{T_{ii} + \sum_{j \neq i} F_{ij}} \tag{2.4}$$

$$\text{PPV}_i = \frac{T_{ii}}{T_{ii} + \sum_{j \neq i} F_{ji}} \tag{2.5}$$

Let N be the total number of samples in the dataset. The *global* sensitivity, PPV, and accuracy of the classifier are calculated as:

$$\text{Sensitivity} = \frac{\sum_{i=1}^{C} \text{Sen}_i}{C} \tag{2.6}$$

$$\text{PPV} = \frac{\sum_{i=1}^{C} \text{PPV}_i}{C} \tag{2.7}$$

$$\text{Accuracy} = \frac{\sum_{i=1}^{C} T_{ii}}{N} \tag{2.8}$$

When $C = 2$, the 2×2 confusion matrix is often reported as *True Positives* (TP), *True Negatives* (TN), *False Positives* (FP), *False Negatives* (FN) and sensitivity, specificity, and accuracy are defined as below.

$$M = \begin{matrix} & \begin{matrix} P & P \\ \text{Positive} & \text{Negative} \end{matrix} & \\ \begin{pmatrix} TP & FN \\ FP & TN \end{pmatrix} & \begin{matrix} \text{Ground Positive} \\ \text{Ground Negative} \end{matrix} \end{matrix} \tag{2.9}$$

$$\text{Sensitivity} = \frac{TP}{TP + FN} \tag{2.10}$$

$$\text{Specificity} = \frac{TN}{TN + FP} \tag{2.11}$$

$$\text{Accuracy} = \frac{TP + TN}{TP + TN + FP + FN} \tag{2.12}$$

$$\text{F1} = \frac{2TP}{(2TP + FP + FN)} \tag{2.13}$$

Besides, other metrics that have been popularly used are false positive rate (*fall-out*), false negative rate (*miss rate*), positive/negative likelihood ratios, and F1-score [47]. F1-score, which is the harmonic mean of precision and sensitivity, has best value at 1 and worst at 0, is calculated as

In the biomedical literature, intra-class correlations (ICCs) [48] has been also used to assess the accuracy performance of classifier (i.e., regarding to the agreement between the classifier and human-labels, called *raters*) [33, 34]. Though ICCs have various forms, this thesis only considers one of six forms of interpretation as described by Shrout and Fleiss in 1979 [48] and developed by McGraw and Wong [31]. Specifically, ICCs are calculated using data from two-way random effect analysis of variance *models* (designation of *ICC(A,1)*) [31]. In this model, raters and *subjects* (i.e., samples to be classified) are random selections from among all possible sources; also raters classify all subjects chosen at random with a known method of rating. The type of ICC computation in this work is absolute agreement with single measures that assesses the comparable classification performance of classifiers.

Let O be a data matrix of size $n \times k$ where k is number of raters and n is number of subjects to be rated. $x_{ij} = \mu + r_i + c_j + e_{ij}$ where $i = 1, \ldots, n$ and $j = 1, \ldots, k$. μ is the population mean for all observations. r_i is the row effect that is random independent and normally distributed with mean 0 and variance σ_r^2. c_j is the column effect that is random independent and normally distributed with mean 0 and variance σ_c^2. e_{ij} is the residual effect that is random independent and normally distributed with mean 0 and variance σ_e^2. The row/column/residual effects, $r_i/c_j/e_{ij}$, are random independent and normally distributed with mean 0 and variance $\sigma_r^2/\sigma_c^2/\sigma_e^2$ respectively. More details could be found [31].

$$O = \begin{pmatrix} x_{11} & \cdots & x_{1j} & \cdots & x_{1k} \\ \vdots & \ddots & \vdots & \cdots & \vdots \\ x_{i1} & \cdots & x_{ij} & \cdots & x_{ik} \\ \vdots & \cdots & \vdots & \ddots & \vdots \\ x_{n1} & \cdots & x_{nj} & \cdots & x_{nk} \end{pmatrix} \begin{matrix} \text{Subject}_1 \\ \vdots \\ \text{Subject}_i \\ \vdots \\ \text{Subject}_n \end{matrix} \tag{2.14}$$

$$\begin{matrix} \text{Rater}_1 & \cdots & \text{Rater}_j & \cdots & \text{Rater}_k \end{matrix}$$

Let's denote the mean squared value for sources of variation to be $\mathrm{MS}_R = k\sigma_r^2 + \sigma_e^2$ for rows, $\mathrm{MS}_C = n\sigma_c^2 + \sigma_e^2$ for columns, and $\mathrm{MS}_E = \sigma_e^2$ for error. The term ICC is computed as:

$$\mathrm{ICC} = \frac{\mathrm{MS}_R - \mathrm{MS}_E}{\mathrm{MS}_R + (k-1)\mathrm{MS}_E + \frac{k(\mathrm{MS}_C - \mathrm{MS}_E)}{n}} \qquad (2.15)$$

2.3 Feature Engineering

In applied machine learning, success depends significantly on the quality of data representation (features) [15]. Basic modules involved in a classification application are illustrated in Fig. 2.1. The process of transforming data into features that are more relevant to the problem, called *feature engineering*, can increase prediction accuracy [59]. Features that are highly correlated with labels can make learning/sorting steps in the classification module easy. Conversely if label classes are a very complex function of the features, it could be impossible to build a good model.

> One may argue that when label classes are a very complex function of the features, a complex classification model such as non-linear classifiers using the kernel trick. From the context of an unsupervised and subject-independent classification application, the approach is not always helpful.

While learners can be largely general-purpose, feature engineering is usually domain-specific [15].

This section describes techniques to automatically select salient features from a large exploratory feature pool (*feature selection*). Redundant and irrelevant features are well known to cause poor accuracy so discarding these features should be the first task. Input features should thus offer a high level of discrimination between classes. Feature selection can be done using a data-driven approach and can be used as a common framework for a wide class of problems.

Feature selection has been applied to several applications such as classification, regression, clustering, association rules and other data mining tasks. This technique sometimes is called *variable* (e.g., [19]) or *attribute* selection (e.g., [22]). In a selection algorithm, depending on the involvement of class information (labels), the technique can be a supervised scheme (e.g., [2, 38, 57]) or unsupervised (e.g., [32]).

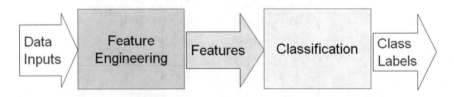

Fig. 2.1 Feature engineering and sorting algorithms involved in a classification task

2.3.1 Feature Relevance

The initial choice of features is often an expression of prior knowledge. Some features may be good representations while others can be irrelevant. Let X be a complete feature set of the data input. $X_i \in X$ is a candidate. $X_i \in X$ is a strongly relevant feature if it contains information that no other candidate does [23]. X_i is a weakly relevant candidate if it has information that also exists in or in conjunction with other ones. Hence, X_i is irrelevant if it is neither strongly relevant nor weakly relevant, otherwise it is relevant.

2.3.2 Feature Selection

Let $S \subset X$ be the desired set of relevant features. $S = s_1, \ldots, s_m$ where m is the number of selected features. The goal of feature selection is to choose S most relevant to the classification task. Identifying the optimal S is an NP-hard problem [6], dependent on a function usually considering size of S, class distribution, accuracy, and relevance [13].

Finding all relevant candidates can be done via an exhaustive search through all the subsets of X, but this is usually not computationally tractable. Starting with an empty set $S = \emptyset$, a forward selection method incrementally adds strongly relevant features but may not find features which are relevant only when combined with others. A backward selection method will start with a full set $S = X$ then remove candidates that are not strongly relevant. This may also discard weakly relevant candidates. The backward elimination can evaluate subsets that contain interacting features thus tends to find better subsets [7]. Search strategies can be an exponential or greedy search [29], or a randomized search [53]. Greedy searches include sequential forward or backward or bi-directional selection techniques. When the space X is very large (e.g., genetic analysis applications), a randomized search is a practical approach in terms of time complexity.

Several interesting examples [23] show that correlated variables may be useless by themselves or strongly relevant ones may be not useful for classification. Thus, a popular approach is finding a *good* subset of the relevant features with a typical process of four steps: subset generation, an evaluation function, a stopping criterion, and validation procedure [13].

After selecting a combination from the space X using a search procedure, each candidate is evaluated and compared according to a certain objective function until a given stopping criterion is satisfied. The evaluation step measures the discriminating ability of a subset with regarding to class labels. Two main groups of objective functions are data intrinsic measures [3] and classifier error rates [14]. The former includes information or uncertainty, distance, and dependence measures. The latter uses the classification accuracy of a classifier involved in predicting the class labels with a selected subset. Automated selection algorithms are grouped into *filter* methods

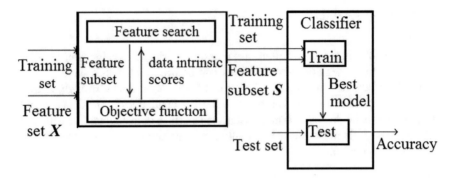

Fig. 2.2 Feature selection model with a filter approach. Data intrinsic measures are often saliency scores using correlation or distance computation

(i.e., use data intrinsic measures and is independent of a classifier) and *wrapper* methods (i.e., utilise a classifier within the algorithm) [25]. Recently, combinations of these methods have been proposed, these being referred to as *hybrid* [13] or *embedded* methods [19].

2.3.3   Automatic Selection Models

2.3.3.1   Filter Models

In *filter* models, S is selected directly from the data and only relies on data intrinsic measures. Feature candidates are ranked by a scoring metric and the highest scoring features are added to S. Figure 2.2 depicts these steps for classification. Many different relevance scores for uncertainty, distance, and dependence measures including Chi-squared, information gain, Pearson correlation, and mutual information [47] have been proposed. For clustering applications, popular scores are Euclidean distance based discrimination [24] and variance ratio [1]. Details of these criteria are discussed in Sect. 3.1.2. Filters select feature subsets independently of the chosen predictor and have low computational cost. Accordingly they may fail to address feature interactions, and the selection criteria are different to the actual learning algorithm.

2.3.3.2   Wrapper Models

A *wrapper* selection approach [25] finds S using a predictive model to score candidates' predictive performance in the evaluation loop. Training instances that are a vector of feature values and a class label are inputs of the learning machine as illustrated in Fig. 2.3. Each new subset is used to train the model and then validated using a different test set. The error of the model is used as the score for the subset.

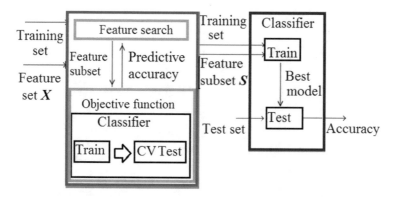

Fig. 2.3 Feature selection model with a wrapper approach [25]. A predictive model is used to score subsets' accuracy performance during cross-validation tests (CV Tests)

To avoid bias, cross-validation (CV) tests [45, 49] that split the training data into traineting set and validation set are used. The validation results are averaged to define the error of the subset during training. Popular methods of CV tests include k-fold CV [35], leave-one-out (LOOCV) [26], random subsampling CV, and bootstrapping CV [16]. The k−fold CV creates k equal sized partitions of the training data. Let N_T be the total number of training samples. Each partition has N_T/k examples. The training loop uses $k − 1$ partitions and validates on the remaining partition, repeating this step k times [35]. Finally, the model is chosen if it has the smallest average validation error. k is often set as 10. LOOCV is a special case of k-fold CV where $k = N_T$. LOOCV is usually practical if N_T is small [26]. The random subsampling CV, a close case of k-fold CV, chooses the validation set as a randomly sampled subset of N_T with a fixed fraction αN_T where$\alpha \in (0, 1)$ and trains with the rest. The loop is also repeated k times. The common settings of the random subsampling CV is $k = 10$ and $\alpha = 0.1$. The bootstrapping CV is a method of random sampling with replacement [16].

A wrapper method directly uses the classification error rate to select the subset, therefore this method tends to perform better than the filter approach. Disadvantages are that it has high computational cost and overfitting may occur.

2.3.3.3 Advanced Models

Hybrid and embedded approaches are *combinations* of the two aforementioned models. In the hybrid method, a filter model is used to reduce the search space for a latter wrapper model [12, 13, 36]. The learning by the wrapper block in the hybrid model could be considered as a *black box* to the work implemented by the filter block [19]. An illustration of this idea is given in Fig. 2.4. Data intrinsic measures are first used to remove non-salient features. Then only top relevant features are evaluated by a classifier of interest. The performance of learning is used to further select the subset that causes the lowest error rate. The hybrid method has advantages of both filter and wrapper approaches, thus it is faster than a normal wrapper and more accurate than

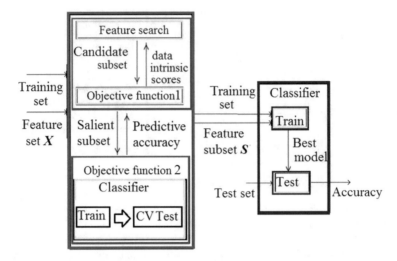

Fig. 2.4 Feature selection model with a hybrid approach. First, data intrinsic measures are used to remove non-salient features. Then high relevant features are evaluated by a classifier (cross-validation, CV, tests)

a filter method. Furthermore, it is less computationally expensive and less prone to overfitting than the wrapper model.

Unlike the hybrid combination, the embedded method finds an optimal S by applying feature weightings during the model building process (Fig. 2.5). An embedded model optimizes feature selection and model together rather than separately in two steps [20]. For example, when using the *Lasso* (Least absolute shrinkage and selection operator) penalty [52], the regularization term of L_1-*norm* is added to the classification error [20] and does feature selection by the sparsity of the Lasso solution. For a given training data set of feature $X \in \mathbb{R}^{n \times d}$ and label Y,

$$X = \begin{bmatrix} x_1(1) & \dots & x_1(d) \\ \vdots & \ddots & \vdots \\ x_n(1) & \dots & x_n(d) \end{bmatrix}, Y = \begin{bmatrix} y_1 \\ \vdots \\ y_n \end{bmatrix} \tag{2.16}$$

The weight w for all feature candidates are considered at once and a regularization penalty that tries to set the weights to zero if their corresponding features are not *relevant* (i.e., the criterion based on the learning objective function) is introduced [20].

$$\min_{w \in \mathbb{R}^d} \frac{1}{n} \sum_{i=1}^{n} (y_i - \langle w, x_i \rangle)^2 + \lambda \|w\|_1, \tag{2.17}$$

where $(y_i - \langle w, x_i \rangle)^2$ is the square loss, λ is a tuning parameter to trade off between loss and penalty, and $\|w\|_1$ is L_1-*norm*. For classification, when λ is sufficiently large, L_1-*norm* will cause most of the weights to be zero [60].

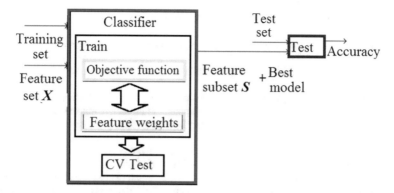

Fig. 2.5 Feature selection model with an embedded approach. Relevance of features are scored by the objective function of the learning as a part of the model building

The embedded method often requires less computing resources than the wrapper methods. However, it is specific to a learning scheme. Sometimes these approaches are called *integrated feature selections*. Algorithms using various learning techniques for an application are referred to as *ensemble* selection methods.

2.4 Summary

This chapter reviewed related background on classification techniques and feature selection. Two common unsupervised classification problems are: anomaly detection which involves two-classs, and x-class sorting which has an unknown number of classes. A confusion matrix is commonly used to evaluate classifier performance. Redundant and irrelevant features are known to cause poor accuracy. It is important to discard such features, and this is often one of the first steps in feature engineering and classification. Automated techniques for feature selection include filter, wrapper, and embedded approaches.

Details of data intrinsic scores used in works of this thesis are discussed in the next Chapter. Specific domain feature extraction is discussed in each scenario of applications (Chap. 4, 5, 6). The next chapters demonstrate algorithms and their theoretical calculations.

References

1. Ackerman M, Ben-David S (2009) Clusterability: a theoretical study. In: International conference on artificial intelligence and statistics, pp 1–8
2. Battiti R (1994) Using mutual information for selecting features in supervised neural net learning. IEEE Trans Neural Netw 5(4):537–550
3. Ben-Bassat M (1982) Pattern recognition and reduction of dimensionality. Handb Stat 2:773–910

4. Blatt M, Wiseman S, Domany E (1996) Superparamagnetic clustering of data. Phys Rev Lett 76:3251–3254
5. Blatt M, Wiseman S, Domany E (1997) Data clustering using a model granular magnet. Neural Comput 9:1805–1842
6. Blum AL, Rivest RL (1992) Training a 3-node neural network is NP-complete. Neural Netw 5(1):117–127
7. Bouchard-Cote A (2009) Lecture notes in feature engineering and selection
8. Capowski JJ (1976) The spike program: a computer system for analysis of neurophysiological action potentials. In: PP Brown (ed) computer technology in neuroscience, (Washington DC: Hemisphere) pp 237–288
9. Chandola V, Banerjee A, Kumar V (2009) Anomaly detection: a survey. ACM Comput. Surv 41(3):15:1–15:58, https://doi.org/10.1145/1541880.1541882
10. Cheeseman P, Kelly J, Self M, Stutz J, Taylor W, Freeman D (1988) Autoclass: a Bayesian classification system. In: Proceedings of the fifth intl workshop on machine learning, pp 54–64
11. Cheng Y (1995) Mean shift, mode seeking, and clustering. IEEE Trans Pattern Anal Mach Intell 17(8):790–799
12. Das S (2001) Filters, wrappers and a boosting-based hybrid for feature selection. ICML, Citeseer 1:74–81
13. Dash M, Liu H (1997) Feature selection for classification. Intell Data Anal 1(3):131–156
14. Doak J (1992) An evaluation of feature selection methods and their application to computer security. University of California, Computer Science
15. Domingos P (2012) A few useful things to know about machine learning. Commun ACM 55(10):78–87
16. Efron B (1983) Estimating the error rate of a prediction rule: improvement on cross-validation. J Am Stat Assoc 78(382):316–331
17. Estivill-Castro V (2002) Why so many clustering algorithms: a position paper. ACM SIGKDD Explor Newsl 4(1):65–75
18. Gibson S, Judy JW, Markovic D (2012) Spike sorting. IEEE Signal Process Mag 29(1):124
19. Guyon I, Elisseeff A (2003) An introduction to variable and feature selection. J Mach Learn Res 3:1157–1182
20. Guyon I, Weston J, Barnhill S, Vapnik V (2002) Gene selection for cancer classification using support vector machines. Mach Learn 46(1–3):389–422
21. Huber PJ (1972) The 1972 wald lecture robust statistics: a review. Ann Math Stat, 1041–1067
22. Jakulin A (2005) Machine learning based on attribute interactions. Fakulteta za racunalništvo in informatiko, Univerza v Ljubljani
23. John GH, Kohavi R, Pfleger K (1994) Irrelevant features and the subset selection problem. In: Machine learning: proceedings of the eleventh international, Morgan Kaufmann, pp 121–129
24. Kira K, Rendell LA (1992) The feature selection problem: Traditional methods and a new algorithm. In: Proceedings of the tenth national conference on artificial intelligence, AAAI Press, AAAI'92, pp 129–134, http://dl.acm.org/citation.cfm?id=1867135.1867155
25. Kohavi R, John GH (1997) Wrappers for feature subset selection. Artif Intell 97(1):273–324
26. Lachenbruch PA, Mickey MR (1968) Estimation of error rates in discriminant analysis. Technometrics 10(1):1–11
27. Leroy AM, Rousseeuw PJ (1987) Robust regression and outlier detection. Wiley series in probability and mathematical statistics 1987:1
28. Lewicki MS (1998) A review of methods for spike sorting: the detection and classification of neural action potentials. Netw Comput Neural Syst 9(4):R53–R78
29. Liu H, Yu L (2005) Toward integrating feature selection algorithms for classification and clustering. IEEE Trans Knowl Data Eng 17(4):491–502
30. Lloyd SP (1982) Least squares quantization in PCM. IEEE Trans Inf Theor 28(2):129–137
31. McGraw KO, Wong SP (1996) Forming inferences about some intraclass correlation coefficients. Psychol Methods 1(1):30
32. Mitra P, Murthy CA, Pal SK (2002) Unsupervised feature selection using feature similarity. IEEE Trans Pattern Anal Mach Intell 24(3):301–312

33. Moore ST, Yungher DA, Morris TR, Dilda V, MacDougall HG, Shine JM, Naismith SL, Lewis SJG (2013) Autonomous identification of freezing of gait in Parkinson's disease from lower-body segmental accelerometry. J Neuroeng Rehabil 10(1):1

34. Morris TR, Cho C, Dilda V, Shine JM, Naismith SL, Lewis SJ, Moore ST (2012) A comparison of clinical and objective measures of freezing of gait in Parkinson's disease. Parkinsonism Relat Disorders 18(5):572–577

35. Mosteller F, Tukey JW (1968) Data analysis, including statistics

36. Ng AY (1998) On feature selection: learning with exponentially many irrevelant features as training examples

37. Oghalai JS, Street W, Rhode WS (1994) A neural network-based spike discriminator. J Neurosci Methods 54(1):9–22

38. Peng H, Long F, Ding C (2005) Feature selection based on mutual information: criteria of max-dependency, max-relevance, and min-redundancy. IEEE Trans Pattern Anal Mach Intell, 1226–1238

39. Pham TT (2011) A real-time neural signal processing system for dragonflies. MS thesis (Advisor Charles M Higgins), Department of Electrical and Computer Engineering, The University of Arizona

40. Pham TT, Fuglevand AJ, McEwan AL, Leong PH (2014) Unsupervised discrimination of motor unit action potentials using spectrograms. In: 36th Annual international conference of the IEEE engineering in medicine and biology society (EMBC) 2014, IEEE, pp 1–4

41. Pham TT, Thamrin C, Robinson PD, McEwan A, Leong PH (2016) Respiratory artefact removal in forced oscillation measurements: a machine learning approach. IEEE Trans Biomed Eng 64(7):1–9

42. Pham TT, Leong PH, Robinson PD, Gutzler T, Jee AS, King GG, Thamrin C (2017a) Automated quality control of forced oscillation measurements: respiratory artifact detection with advanced feature extraction. J Appl Physiol 123(4):781–789

43. Pham TT, Moore ST, Lewis SJG, Nguyen DN, Dutkiewicz E, Fuglevand AJ, McEwan AL, Leong PH (2017b) Freezing of gait detection in Parkinson's disease: a subject-independent detector using anomaly scores. IEEE Trans Biomed Eng 64(11):2719–2728

44. Potts RB (1952) Some generalized order-disorder transformations. Math Proc Camb Philos Soc 48(1):106109, https://doi.org/10.1017/S0305004100027419

45. Quenouille MH (1949) Approximate tests of correlation in time-series 3. Mathematical proceedings of the cambridge philosophical society 45:483–484

46. Quiroga R, Nadasdy Z, Ben-Shaul Y (2004) Unsupervised spike detection and sorting with wavelets and superparamagnetic clustering. Neural Comput 16:1661–1687

47. Rijsbergen CJV (1979) Inf Retr, 2nd edn. Butterworth-Heinemann, Newton, MA, USA

48. Shrout PE, Fleiss JL (1979) Intraclass correlations: uses in assessing rater reliability. Psychol Bull 86(2):420

49. Stone M (1974) Cross-validatory choice and assessment of statistical predictions. J Royal Stat Soc Ser B (Methodological) 36(2):111–147, http://www.jstor.org/stable/2984809

50. Takahashi S, Anzai Y, Sakurai Y (2003) A new approach to spike sorting for multi-neuronal activities recorded with a tetrode-how ICA can be practical. Neurosci Res 46(3):265–272

51. Teng HS, Chen K, Lu SC (1990) Adaptive real-time anomaly detection using inductively generated sequential patterns. In: 1990 Proceedings of the IEEE computer society symposium on research in security and privacy, IEEE, pp 278–284

52. Tibshirani R (1996) Regression shrinkage and selection via the lasso. J Royal Stat Soc Ser B (Methodological) pp 267–288

53. Vafaie H, De Jong K (1995) Genetic algorithms as a tool for restructuring feature space representations. In: 1995 Proceedings of the seventh international conference on tools with artificial intelligence, IEEE, pp 8–11

54. Vargas-Irwin C, Donoghue JP (2007) Automated spike sorting using density grid contour clustering and subtractive waveform decomposition. J Neurosci Methods 164(1):1–18

55. Wolff U (1989) Comparison between cluster Monte Carlo algorithms in the Ising model. Phys Let B 228:379–382

56. Yang Z, Zhao Q, Liu W (2009) Improving spike separation using waveform derivatives. J Neural Eng 6:046,006–046,018
57. Yu L, Liu H (2004) Efficient feature selection via analysis of relevance and redundancy. J Mach Learn Res 5:1205–1224
58. Zhang C, Zhang X, Zhang MQ, Li Y (2007) Neighbor number, valley seeking and clustering. Pattern Recogn Lett 28(2):173–180
59. Zhang C, Kumar A, Ré C (2014) Materialization optimizations for feature selection workloads. In: Proceedings of the 2014 ACM SIGMOD international conference on management of data, ACM, pp 265–276
60. Zhu J, Rosset S, Hastie T, Tibshirani R (2004) 1-norm support vector machines. Adv Neural Inf Process Syst 16(1):49–56

Chapter 3
Algorithms

3.1 Feature Engineering

Discarding redundant and irrelevant features which are well known to cause low learning performance are first tasks of feature engineering. In this thesis, a hybrid model is proposed to utilize advantages of both filter and wrapper approaches. The criterion of accuracy is evaluated using simple/low computational cost classification algorithms. Several saliency criteria have been used at one time and also a voting-based process is suggested to improve the robustness of features across parameter settings.

3.1.1 Automated Selection Process

Given a large exploratory feature pool, a voting process is suggested to select the best feature [13, 15]. This process uses three levels of selection: saliency, robustness, and accuracy; called *Round1, Round2, Round3* respectively (Fig. 3.1). After each level, selected candidates become more favourable. Specifically, *Round1* suggests the most salient and discriminative subset using mutual information (MI), separability calculated using Euclidean distances (DIS), and the variance ratio of clusters (VarRatio). After identifying a highly salient subset based on relevance scores, *Round2* examines if the candidates are robust across window sizes or criteria (i.e., are shared in more than one list). Finally, given a proposed detector, *Round3* tests the detection performance when applying these features.

© Springer Nature Switzerland AG 2019
T. T. Pham, *Applying Machine Learning for Automated Classification of Biomedical Data in Subject-Independent Settings*, Springer Theses, https://doi.org/10.1007/978-3-319-98675-3_3

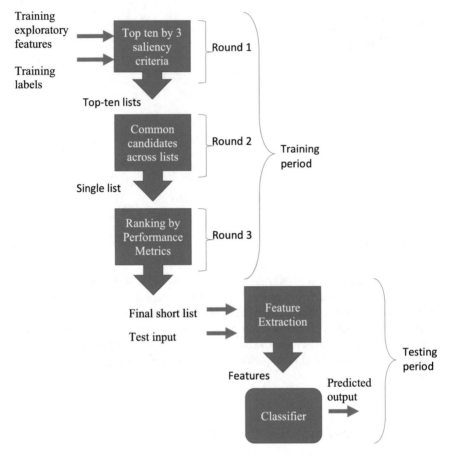

Fig. 3.1 Feature selection process with three phases Feature selection process with three phases: saliency, robustness, and accuracy; called *Round1, Round2, Round3*

3.1.2 Saliency Criteria

Saliency criteria are conditions used to select the best subset of features. Salient features are candidates that are more relevant to the target variable of the classification task and/or more discriminative. There are three different ranking scores for each feature. The mutual information (*MI*) between a candidate and the class label measures a correlation or the relevance of the feature. The other two types of scores (*DIS* and *VarRatio*) assess the separability of a candidate's values across class labels, i.e., clusterability. *DIS* scores present the relationship of Euclidean distances between clusters while the *VarRatio* uses variance ratio of clusters.

3.1.2.1 Mutual Information

Let X be a discrete random variable $X \in \mathbb{X}$ and C be a target variable ($c \in \mathbb{C}$) where $\mathbb{X}$ is the input set and $\mathbb{C}$ is the output set, i.e., class label set. The *entropy* $H_b(X)$ of X measures its uncertainty [16]. $H_b(X)$ is computed as in Eq. (3.2). The conditional entropy of X given C is defined by:

$$H(X|C) = - \sum_{c \in \mathbb{C}} p(c) \sum_{x \in \mathbb{X}} p(x|c) \log p(x|c) \tag{3.1}$$

where

$$H_b(X) = - \sum_{x \in \mathbb{X}} p(x) \log_b p(x), \tag{3.2}$$

and b is the base of the logarithm. In this work, $b = 2$ and hence the entropy is in bits. $p(x) = P\{X = x\}, x \in \mathbb{X}$ is probability mass function.

The mutual information between X and C, $MI(X; C)$, measures the amount of information shared by X and C, i.e., the relevance of X to C, (Eq. (3.3)) [16].

$$MI(X; C) \stackrel{\text{def}}{=} H(X) - H(X|C)$$
$$= \sum_{x \in \mathbb{X}} \sum_{c \in \mathbb{C}} p(x, c) \log \frac{p(x, c)}{p(x)p(c)} \tag{3.3}$$

3.1.2.2 Clusterability

To assess discrimination of features, relevant candidates are considered having nearest instances (by Euclidean distances) of same class closer and having nearest ones of other classes more far apart. The weighting of these distances, called DIS, is calculated with the *RELIEF* algorithm [7] as implemented in an existing package [3] and built-in packages by MATLAB (The MathWorks Inc., Natick, MA, 2000).

For VarRatio scores, the variance ratio of a feature X is the ratio of variance calculation for data within a class (i.e., a cluster) and data between classes (i.e., clusters) and is defined as in Eq. (3.4). A higher *VarRatio(X)* implies that it is easier to cluster X [1], therefore the feature is more desirable.

$$VarRatio(X) \stackrel{\text{def}}{=} \frac{B_C(X)}{W_C(X)} \tag{3.4}$$

where $B_C(X)$ is the between-cluster variance and $W_C(X)$ is the within-cluster variance.

3.2 Classification Algorithms

The aforementioned feature engineering process can be described for several applications, classification scenarios are conversely better illustrated with particular applications. This thesis includes three types of applications for the number of class in
a classification task. Two-class with data points or collective instances are first two
discussed, then an unknown class number is presented. For the data point case, data
recording from accelerometers to recognize the timing of a special abnormal walking status in patients with Parkinson's disease that is so-called *freezing of gait (FoG)*
is considered (details of data and application are in Chap. 4). Each data point is an
instance of acceleration data in a timeseries. FoG instances can be detected as anomalies. For the collective case, data from each breath cylce during lung function tests
form an instance to classify. Respiratory artefactual breaths can also be detected as
collective anomalies. For the last case, action potentials during electrophysiological
recordings are typical inputs for a sorting problem where the class number can be
unknown.

3.2.1 Point Anomaly Detection Scheme

Let $\phi(n)$ be a value of feature vector of length N at time n. For example, in freezing
of gait (FoG) detection, they are often energy metrics of acceleration data [2, 4–6,
8–10, 17] Because there has been also a contextual abnormal increasing of energy
in a specific frequency band of data, an anomaly score that has been shown in the
preliminary work of this thesis to be efficient for this detection, called $A(n)$, of $\phi(n)$
at time n. $A(n)$ determines if the feature value $\phi(n)$ extracted from a window at time
n is higher than a threshold (Eq. (3.5)).

$$A(n) = \text{sign}\left(\phi(n) - \frac{\alpha}{|n-1|}\sum_{m=1}^{n-1}[\phi(m)A(m)]\right) \qquad (3.5)$$

where $A(1)$=1, $n \in [2, N]$, $\alpha > 0$ is a scale factor, and $\text{sign}(x)$ is 1 if $x > 0$ else 0.
    This thesis proposes an ASD (Anomaly Score Detector) using $A(n)$ to detect such
point anomalies in the FoG application.

3.2.2 Collective Anomaly Detection Scheme

An anomalous collection of related adjacent datan instances is referred to as a collective anomaly. In this thesis, one example for this real life anomalies is detecting respiratory artefact in lung function tests using complete-breath approaches.

Each sample contains consecutive time instances corresponding to the beginning and ending of a respiratory cycle. An artefactual breath is an anomaly if it does not conform to the expected behaviour as normal breaths (e.g., contaminated by a negative respiratory resistance value). Earlier works [13, 14] suggest that a binary anomaly score can be used to detect the respiratory artefacts. Via thresholding, a breath is marked as an artefact and discarded if one of any data points in the cycle has its features exceed a given upper bound or are less than a lower one.

Given a set of breaths, $\mathbb{B}$, let $\phi_b(n)$ be a value of the nth instance of feature vector for breath b ($b \in \mathbb{B}, n \in [1, N]$ where N is the length of b). In the respiratory artefact removal application, $\phi_b(n)$ often includes numerical information about the average respiratory resistance, volume, and flow of the breath. The anomaly score, called A_b, of ϕ_b determines if the feature is within a limited range of $[\theta_L, \theta_H]$ $\forall n \in [1, N]$ (Eq. (3.6)).

$$A_b = \prod_{n=1}^{N} \text{sign}\left(\phi_b(n) - \theta_L\right) \text{sign}\left(\theta_H - \phi_b(n)\right) \qquad (3.6)$$

where θ_L and θ_H are defined in a particular application, and $\text{sign}(x)$ is 1 if $x > 0$ else 0.

One example of applying Eq. (3.6) is discussed further in Chap. 5 with quartile information extracted from measurements. Though calculation of θ_L and θ_H can be different across features, to simplify parameter settings, a similar computation is illustrated with real data in this work.

3.2.3 Correlation Based Spike Sorting Scheme

In the third scenario of applied machine learning for automated classification in biomedical data, an unknown-class sorting application is presented. For example, motor unit (MU) activity is analysed using intramuscular EMG data. MU action potential (MUAP or so-called *spike*) waveforms can be classified into MU groups they belong to based on MUAP morphology (*Spike sorting*). The number of MU is not known; i.e., the number of classes in the classification task is not given.

Inspired by a preliminary work of this thesis [11], the correlation between spikes can be used as the similarity measure for such clustering applications [12]. In this work, after a step of feature extraction from MUAPs (details as in Sect. 6.3), each spike is presented by a feature vector. Let X and Y be two feature vectors of $MUAP_X$ and $MUAP_Y$, respectively, and $r_{X,Y}$ be the correlation between X and Y. Equation 3.7 depicts the calculation of Pearson's correlation by definition for a range of $[-1, 1]$. A high $r_{X,Y}$ indicates high similarity between X and Y, the aforementioned thresholding method can also be applied for this application. Given threshold parameter Θ_{co}, a spike Y belongs to the same class with X if the correlation $r_{X,Y} > \Theta_{co}$.

$$r_{X,Y} = \frac{\mathscr{C}\{X, Y\}}{\sigma_X \sigma_Y} \tag{3.7}$$

where $r_{X,Y}$ is the correlation coefficient between MUAP X and MUAP Y. $\mathscr{C}\{X, Y\}$ is the covariance of two feature vectors X and Y. σ_X and σ_Y are the variances of X and Y, respectively.

Let X be a set of spikes and c_i be a class assignment variable of a spike s_i ($s \in S$). Initially, there is only one single class, c_1, that contains the very first spike waveform collected ($i = 1$). Then the class assignment c_j of a spike s_j is determined by $c_j = c_i$ if $r_{s_i,s_j} > \Theta_{co}$ where Θ_{co} is a parameter of the application, i and j are indices of spikes (initially $i = 1$). Let R be the remaining set of spikes with undefined c_j after the above process. These assignments are repeated in a loop until R is empty.

3.3 Summary

Methods of automated feature selection used in this thesis are described in the first two sections. Then the proposed classification techniquesfor applications are described in the next three sections. The main goal of this task is grouping data into distinct subsets. In each section, specific anomaly detection and spike sorting algorithms are introduced.

1. Three saliency scores chosen are mutual information (MI), separability calculated using Euclidean distances (DIS), and the variance ratio of clusters (VarRatio).
2. Top ranking candidates using these criteria are selected as relevant subsets.
3. The selected features are further filtered using robustness voting criterion across parameter settings (e.g., window sizes).
4. Finally, performance metrics are used to select the final proposed feature set for classification applications.
5. For data point and collective sample detection, anomaly scores at time instances represent if the feature values extracted from windows exceed a threshold.
6. For unsupervised spike sorting that has no pre-defined class information, high correlation coefficients indicate high similarity between spikes. If the similar degree exceeds a given threshold, the spikes are considered to belong to the same class.

The next chapters will discuss in detail for each scenario compared with domain-knowledge feature extraction.

References

1. Ackerman M, Ben-David S (2009) Clusterability: a theoretical study. In: International conference on artificial intelligence and statistics, pp 1–8
2. Bachlin M, Plotnik M, Roggen D, Maidan I, Hausdorff J, Giladi N, Troster G (2010) Wearable assistant for Parkinson's disease patients with the freezing of gait symptom. IEEE Trans Inf Technol Biomed 14(2):436–446
3. Brown G, Pocock A, Zhao MJ, Luján M (2012) Conditional likelihood maximisation: a unifying framework for information theoretic feature selection. J Mach Learn Res 13(1):27–66
4. Cole B, Roy S, Nawab S (2011) Detecting freezing-of-gait during unscripted and unconstrained activity. In: Annual international conference of the IEEE engineering in medicine and biology society. EMBC, pp 5649–5652
5. Gazit E, Bernad-Elazari H, Moore S, Cho C, Kubota K, Vincent L, Cohen S, Reitblat L, Fixler N, Mirelman A et al (2015) Assessment of Parkinsonian motor symptoms using a continuously worn smartwatch: preliminary experience. Mov Disord 30:S272–S272
6. Han J, Lee W, Ahn T, Jeon B, Park KS (2003) Gait analysis for freezing detection in patients with movement disorder using three dimensional acceleration system. In: Proceedings of the 25th Annual international conference of the ieee engineering in medicine and biology society, vol 2, pp 1863–1865
7. Kira K, Rendell LA (1992) The feature selection problem: traditional methods and a new algorithm. In: Proceedings of the Tenth national conference on artificial intelligence, AAAI Press, AAAI'92, pp 129–134. http://dl.acm.org/citation.cfm?id=1867135.1867155
8. Mazilu S, Hardegger M, Zhu Z, Roggen D, Troster G, Plotnik M, Hausdorff J (2012) Online detection of freezing of gait with smartphones and machine learning techniques. In: 6th International conference on pervasive computing technologies for healthcare (PervasiveHealth), pp 123–130
9. Moore S, MacDougall H, Ondo W (2008) Ambulatory monitoring of freezing of gait in Parkinson's disease. J Neurosci Methods 167(2):340–348
10. Moore ST, Yungher DA, Morris TR, Dilda V, MacDougall HG, Shine JM, Naismith SL, Lewis SJG (2013) Autonomous identification of freezing of gait in Parkinson's disease from lower-body segmental accelerometry. J Neuroeng Rehabil 10(1):1
11. Pham TT, Higgins CM (2014) A visual motion detecting module for dragonfly-controlled robots. In: 36th Annual international conference of the IEEE engineering in medicine and biology society (EMBC) 2014. IEEE, pp 1666–1669
12. Pham TT, Fuglevand AJ, McEwan AL, Leong PH (2014) Unsupervised discrimination of motor unit action potentials using spectrograms. In: 36th Annual international conference of the IEEE engineering in medicine and biology society (EMBC) 2014. IEEE, pp 1–4
13. Pham TT, Thamrin C, Robinson PD, McEwan A, Leong PH (2016) Respiratory artefact removal in forced oscillation measurements: a machine learning approach. IEEE Trans Biomed Eng 64(7):1–9
14. Pham TT, Leong PH, Robinson PD, Gutzler T, Jee AS, King GG, Thamrin C (2017a) Automated quality control of forced oscillation measurements: respiratory artifact detection with advanced feature extraction. J Appl Physiol 123(4):781–789
15. Pham TT, Moore ST, Lewis SJG, Nguyen DN, Dutkiewicz E, Fuglevand AJ, McEwan AL, Leong PH (2017b) Freezing of gait detection in Parkinson's disease: a subject-independent detector using anomaly scores. IEEE Trans Biomed Eng 64(11):2719–2728
16. Shannon C (1948) A mathematical theory of communication. Bell Syst Tech J 27(3):379–423
17. Zach H, Janssen AM, Snijders AH, Delval A, Ferraye MU, Auff E, Weerdesteyn V, Bloem BR, Nonnekes J (2015) Identifying freezing of gait in Parkinson's disease during freezing provoking tasks using waist-mounted accelerometry. Parkinsonism Relat Disord 21(11):1362–1366

Chapter 4
Point Anomaly Detection: Application to Freezing of Gait Monitoring

4.1 Background on Gait Freezing Detection in Parkinson's Disease

Gait is one of the most affected motor characteristics of Parkinson's disease (PD). Freezing of gait (FoG) that is defined as a motor block of movement, especially before gait initiation, during turns or when meeting obstacles [3] is one of the most common symptoms (e.g., forty-seven percent of the patients reported experiencing freezing regularly [19]). Moreover, there is a strong relationship between FoG and falls in people with PD [3, 18, 27].

Current clinical FoG assessment methods are self-reported diaries from patients (e.g. the Unified Parkinson's Disease Rating Scale (UPDRS) [7], Freezing of Gait Questionnaire [11]) and manual video analysis of walking tasks [25, 37]. These methods are subjective. UPDRS has poor agreement with expert labels (the kappa statistic only ranged from 0.49 to 0.78) [29]. The reliability of existing manual video assessment is not robust (within or across multiple participant recruitment sites); the intra-rater reliability is remarkably low [26]. An additional difficulty lies in provoking FoG during routine clinical examinations [31].

Objective FoG detection is very much desirable, especially out-of-lab deployment with wearable devices [9, 23]. Compared with kinematic and electrophysiological data (e.g. electromyographic and electroencephalogram), acceleration data have been widely adopted thanks to the small size of accelerometers, making them suitable for wearable systems. An early effort was reported [13] with two accelerometers at both ankles. Han et al. [13] found that freezing gait has high frequency components ($6 \rightarrow 8$ Hz) compared with normal gait (2 Hz). Wavelet analysis [6] has been used to classify normal and freezing gait (including the ratios of each level's power to discriminate the freezing and resting states) [13]. A freezing index (FI), defined as the power in the *freeze* band ($3 \rightarrow 8$ Hz) divided by the power in the *locomotor* band ($0.5 \rightarrow 3$ Hz) [23], has been used to build FoG detectors [2, 5, 9, 20, 23, 24, 38]. From a machine learning perspective, two main

© Springer Nature Switzerland AG 2019
T. T. Pham, *Applying Machine Learning for Automated Classification of Biomedical Data in Subject-Independent Settings*, Springer Theses, https://doi.org/10.1007/978-3-319-98675-3_4

classification approaches are: simple thresholding techniques [2, 9, 23, 24, 38] and supervised/semi-supervised learning classifiers [5, 20]. However, these reports were based on separate channels.

To extract features, two types of inputs can be used: single input (e.g., single channels from single sensors (SCSS), the sum of squares of all three channels of single sensors (MCSS)) and multiple inputs (i.e., multiple channels of multiple sensors, MCMS). While SCSS and MCSS have been well studied, MCMS is considered for the first time in this work. Note that the work [24] examined one case of using seven sensors (only single axis from each sensor was used) that was the majority votes of seven outputs which we categorize into the SCSS group. MCMS to is used to refer to a case where feature values are computed from a matrix of inputs.

Recently, apart from *FI*, several features from accelerometer data (e.g., average, standard deviation, variance, median, entropy, energy, and power) have been proposed for FoG detectors [2, 5, 9, 13, 20, 21, 23, 24, 38]. Advanced statistical techniques to assess gait of human in general (e.g., postural control) can be found in a comprehensive feature investigation [33], however the work was concerned with 3D motion analysis for trajectory data using a single accelerometer at the lumbar. The authors concluded that no measure in their study was able to discriminate the gait patterns of individuals within clinical groups of PD and peripheral neuropathy. Furthermore, freezing of gait data was not collected in that study.

On the other hand, we explore the new combinations of inputs. We investigate three new computation methods: the spectral coherence [4], *multi-channel FI* (FI_{MC}), and Koopman spectral analysis [17] (FI_K). FI_{MC} and FI_K, are applicable only to MCMS inputs.

With regard to feature selection algorithms, several other features were compared with FI [21] including statistical and zero crossing rate (SCSS group), sum of the Euclidean norm of magnitude, eigenvalues of the covariance matrix, the mean energy, and principal component analysis over the three axes of the sensor (MCSS group). Nevertheless, the report [21] was solely based on mutual information (MI) that measures the correlation of features with labels (Shannon's information theory) [34]. This selection could not guarantee the clusterability [1] of the selected features. A classifier will have better performance with more discriminative features. Thus, in our work, we explore several extra new features that are extracted from new analysing function or from multiple channels/sensors concurrently and create an *exploratory* feature pool. Besides MI, we rank the pool using two additional saliency criteria: the variance ratio of clusters [1] and the separability calculated by Euclidean distances from an instance to a *near-hit* and *near-miss* [16].

Several works have been proposed recently with moderate subject-independent results. For example, despite using the same channel (the vertical axis of the ankle sensor), a *global threshold FI* of 2.3 with 6*s* windowing was suggested [23], then later on another global FI of 3 with 7*s* windowing was reported [24]. By examining the same three locations of sensors as before [24], Zach et al. [38] made a different choice for the global FI, namely 1.4 (2*s* windows and the dorsoventral direction of the lumbar sensor). In the mean time, the model learning based classifiers have been only optimized for subject-dependent or group-dependent settings [5, 20].

A primary reason hindering subject-independent performance lies in the generalization of parameters. One example could be a strong *context* dependence of parameters in conjunction with large subject-variability [24]. Therefore, we hypothesize that a detector based on anomaly scores (called *ASD*) can improve significantly the subject-independent performance.

4.2 FoG Detector

A data point, i.e., a time instance, is a *point anomaly* if its behaviours differs from other data points. For example, in freezing of gait (FoG) detection applications for patients with Parkinson's disease, the time instance a patient suffers from gait freezing is of interest and is used to compute how long a freezing event lasts for. Therefore, FoG instances can be considered point anomalies against normal behaviour of the patient (i.e., non-FoG). In this application, features of an instance are often extracted in a sliding window manner. Using anomaly detection techniques, FoG events can be detected by anomaly scores [28].

ASD employs an adaptive rather than fixed threshold. Inspired by observations of an increase in FI during a FoG event (versus a locomotor activity) [13, 23], we investigate if this is also the case with other features. When the current feature value of a data window is lower than the on-the-fly threshold, we consider the window a *potential* non-FoG epoch. During detection, the threshold at a time is the average of all previous values from *potential* non-FoG epochs (initially is the first data window). Thus, ASD only engages in learning from recent activity periods of the corresponding subject rather than learning globally from several seen subjects. The initial delay of learning is one window size (e.g., $2s$). Furthermore, ASD can avoid any effect of diurnal variation. After a given typical medium clinical trial duration (e.g., about thirty minutes in our datasets), we reset the on-the-fly threshold of ASD, i.e., it may not be averaged across *activity contexts*. If the reset happens at a anomaly instance, the low pass filter effect of the ASD (Sect. 3) eventually converges to a normal value. In other words, ASD is inherently independent of subject variability and diurnal variation.

This thesis demonstrates an example of ASD using $A(n)$ (Sect. 3.2.1). Initially, the first data window is assumed to be normal behaviour. If this assumption is wrong, the averaging effect of Eq. (3.5) is expected to low pass filter FoG events and eventually converges to a normal value. The results of a setting with $\alpha = 1$ (i.e no scaling deviation are reported in this work.

4.3   Data Set

We first developed our algorithm with a dataset from the Daphnet project [2]. Then we deployed out-of-sample tests with a different dataset collected independently as one part of a larger project for FoG studies [35]. FoG annotations/labels were assessed on the Movement Disorder Society Unified Parkinson Disease Rating Scale Section III (MDS-UPDRS-III) [12] and Hoehn and Yahr stage score [14].

4.3.1   Development Set

Seven male and three female advanced PD patients who could walk unassisted in the OFF period were recruited at Tel Aviv Sourasky Medical Center (TASMC) in Israel as a part of the EU FP6 Daphnet project (a collaboration with ETH Zurich, Switzerland) [2]. These ten participants (66.5 ± 4.8 years old) have been diagnosed with PD for 13.7 ± 9.67 years (Hoehn and Yahr score [14] (H&Y) is 2.6 ± 0.65). The dataset was recorded in the lab during the OFF stage of the medication cycle of the participants, except for two participants who reported a frequent FoG experience during the ON stage. As illustrated in Fig. 4.1, three tri-axial (x—anterior/posterior, y—medial/lateral, z—vertical) accelerometers were attached at the shank (above the

Fig. 4.1   Three tri-axial accelerometers were attached at the shank, thigh, and lower back

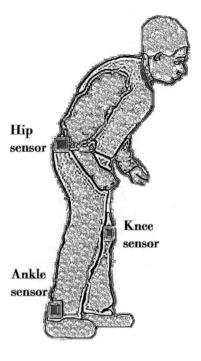

Hip sensor

Knee sensor

Ankle sensor

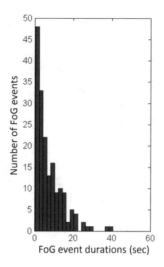

Fig. 4.2 Histogram of FoG episode durations (second) according to the annotations

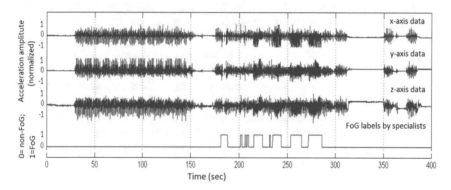

Fig. 4.3 Three data channels for one sample recording from a participant

ankle), thigh (above the knee), and lower back (trunk, above the hip) using elasticized straps. Data was recorded at 64 Hz and transmitted via a Bluetooth link. Figure 4.3 illustrates three data channels for one sample recording from a participant.

Three walking tasks (10−15 min each) were conducted: walking a straight line, with numerous turns, and a daily living activity (e.g., fetching coffee, opening doors); more details as in [2]. Three tri-axial accelerometers were attached at the shank, thigh, and lower back using elasticized straps. Annotation and simultaneous video taping were used by physiotherapists to determine the start/end times of FoG episodes. A FoG event label started when the gait pattern (i.e., alternating left – right stepping) was arrested and ended when the pattern was resumed [2]. The study was approved by the local Human Subjects Review Committee, and was performed in accordance with the ethical standards of the Declaration of Helsinki [2].

This dataset is recommended to benchmark automatic methods for gait freezing detection from wearable acceleration sensors. A total of five hundred minutes of data were collected. Eight participants had FoG while two did not. The walking distance and number of turns depended on each participant's execution. A total of 237 freezing events ($0 \rightarrow 66$ per subject, 23.7 ± 20.7) were recognized using video analysis by physiotherapists (Fig. 4.2). This is used as the *ground truth* in our accuracy evaluations. For algorithm development (i.e., ranking features and tuning parameters), this work uses a random sample of 70% (five) participants who had FoG events (66 ± 5.9 years old, with PD for 16.2 ± 10.15 years, H&Y score: 2.3 ± 0.44). For out-of-sample tests, this work uses the remaining subjects (66.8 ± 4.1 years old, with PD for 11.2 ± 9.6 years, H&Y score: 2.9 ± 0.74). Specifically, the test set consists of 30% (three) of participants who had FoG and the others with no FoG.

4.3.2   Test Set

We employed an independent data set for out-of-sample tests from a larger FoG study project [35]. This set included 24 patients (mean $\pm$SD age: 69 ± 8.41 with advanced PD (mean $\pm$SD Hoehn and Yahr: 2.66 ± 0.53; UPDRS III: 40.24 ± 11.06) at Parkinsons Disease Research Clinic (the Brain and Mind Research Institute, University of Sydney, NSW Australia). These participants had severe self-reported freezing behavior and satisfied UKPDS Brain Bank criteria [10]. The subjects were deemed unlikely to have dementia or major depression according to DSM-IV criteria (by consensus rating of a neurologist and a neuropsychologist) and had a mean $\pm$SD Mini-Mental State Examination (MMSE) [8] score of 28.57 ± 1.61. The study was approved by the Human Research and Ethics Committee at the University of Sydney and written consents from participants obtained.

Participants were recorded in the practically-defined off state following overnight withdrawal of dopaminergic therapy. Six patients also had Deep Brain Stimulation (five Subthalamic Nuclei and one Pedunculopontine Nuclei), which were turned off for one hour prior to assessment. None of the patients described any increase in freezing behavior following the administration of their usual dopaminergic therapy.

Walking tasks were described in detail [35] that were designed to best provoke FoG during data collection. Participants started from a sitting position, walked along a corridor about five meters meeting a marked square on the floor (size of 0.6 m) then made a turn ($180°$ or $540°$ to the left or right of the subject) as shown in Fig. 4.4. Each task was introduced to a participant at the beginning of the trial, if the subject had failed to meet the procedure, the measurement was abandoned. Each trial was started by a signal from the investigator and was completed on return to the beginning position.

Data from accelerometers were acquired by seven tri-axial sensors attached to each subject at the back, foot, thigh and/or knee (further details as in the previous work [24]). These sensors were inertial measurement units (IMUs—Xsens MTx, Enschede, Netherlands) that were $38 \times 53 \times 21$ mm and 30 g. Data was transmitted

Fig. 4.4 Walking tasks description for FoG detection. Tasks started from a sitting position, walked along a corridor about five meters meeting a marked square on the floor (size of 0.6 m) then made a turn (180° or 540° to the left or right of the subject) [35]

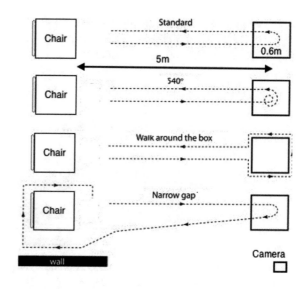

via a wireless link to a computer (sampling frequency of 50 Hz). Clocks of computer for data acquisition and of the video camera were used to synchronize the timing between clinical annotations and acceleration measurement.

Manual assessment of FoG was made by clinicians (neurologist/neuropsychologist experienced in FoG) using video taped during each trial. These annotations were converted to binary labels ("0" for non-FoG or "1" for FoG each time instance). Each trial was assessed by two clinicians. The official label was determined hlto be FoG if at least one clinician marked as such. Agreement of these two raters was previously reportedd with high intraclass correlation coefficient (0.82 for number of FoG epochs and 0.99 for percent time frozen) [24, 35]).

For a better comparison with the development stage, we selected data from all three tri-axial channels at three sensor locations of back, left thigh, and left shank. There were total of 71 trials across 15 subjects with six different walking procedures.

4.4 Feature Extraction

4.4.1 Existing Features

Bachlin et al. [2] had reported a relationship between FoG status and the power spectral density distribution from $0 \rightarrow 128$ Hz for walking, FoG, and standing. According to the published distribution [2], $0 \rightarrow 30$ Hz is the main frequency range for human movement. Walking and FoG status have about 96% of the total energy while the PSD of standing is dominated by sensor noise. They also claimed that the total energy content of standing is substantially lower than for FoG or walking. Hence, this report applied these special features of *LocoBand* of [0.5 → 3 Hz] and *FreezeBand* of [3 → 8 Hz] to help detect a FoG event with power and freeze index values P (Eq. 4.1) and *FI* (Eq. 4.2) in this work.

$$P = P_H + P_L \tag{4.1}$$

$$FI = \frac{P_H}{P_L} \tag{4.2}$$

$$P_H = \frac{\sum_{i=H_1+1}^{H_2} P_{XX}(i) + \sum_{i=H_1}^{H_2-1} P_{XX}(i)}{2fs} \tag{4.3}$$

$$P_L = \frac{\sum_{i=L+1}^{H_1} P_{XX}(i) + \sum_{i=L}^{H_1-1} P_{XX}(i)}{2fs} \tag{4.4}$$

where P_{XX} is power spectrum of acceleration data; N_{FFT} is the window size of FFT transform; fs is sampling frequency, $H_1 = \frac{3N_{FFT}}{fs}$, $H_2 = \frac{8N_{FFT}}{fs}$, $L = \frac{0.5N_{FFT}}{fs}$.

4.4.2   New Features

We study four new features. The first two use single input data channels: the maximum and number of peaks in the spectral coherence [4] (called C_{XYNpks} and C_{XYmax}). The others use multiple inputs: FI_{MC} and FI_K. Let x and y be two consecutive data windows. The spectral coherence C_{XY} between x and y using the Welch method [4] is $C_{XY}(\omega) = \frac{P_{XY}(\omega)}{\sqrt{P_{XX}(\omega).P_{YY}(\omega)}}$ where ω is frequency, $P_{XX}(\omega)$ is the power spectrum of signal x, $P_{YY}(\omega)$ is the power spectrum of signal y, and $P_{XY}(\omega)$ is the cross-power spectrum for signals x and y. When $P_{XX}(\omega) = 0$ or $P_{YY}(\omega) = 0$, then $P_{XY}(\omega) = 0$ and $C_{XY}(\omega)$ is assumed as zero. The power and cross spectra are: $P_{YY}(\omega) = \mathfrak{F}_y(\omega).\overline{\mathfrak{F}_y(\omega)}$; and $P_{XY}(\omega) = \mathfrak{F}_x(\omega).\overline{\mathfrak{F}_y(\omega)}$.

Let a matrix $\mathbf{X}$ of size $N \times M$ represent a N-channel recording session with M regularly spaced time samples. Similar to the single input FI, FI_{MC} is the ratio of powers P_H to P_L (i.e., for the *freeze* and *locomotor* bands) that are summations of single powers over N channels. Specifically:

$$P_H = \frac{1}{2f_s} \sum_{n=1}^{N} [\sum_{i=H_1+1}^{H_2} [P_{XX_n}(i)] + \sum_{i=H_1}^{H_2-1} [P_{XX_n}(i)]] \tag{4.5}$$

$$P_L = \frac{1}{2f_s} \sum_{n=1}^{N} [\sum_{i=L+1}^{H_1} [P_{XX_n}(i)] + \sum_{i=L}^{H_1-1} [P_{XX_n}(i)]] \tag{4.6}$$

$$FI_{MC} = \frac{P_H}{P_L} \tag{4.7}$$

where N is number of channels, n is the channel identification, f_s is sampling frequency, $H_1 = \frac{3N_{FFT}}{f_s}$, $H_2 = \frac{8N_{FFT}}{f_s}$, $L = \frac{0.5N_{FFT}}{f_s}$.

We also extract another type of freeze index from **X**, called FI_K, that results from a spectral analysis using the Koopman operator [17]. This computation was introduced to study the spectrum of Hamiltonian systems by using linear transformations on Hilbert space. Dynamic Mode Decomposition [32] is a technique to estimate a linear model with Koopman eigenfunctions and eigenvalues. Inspired by a feature extraction application [15], Koopman eigenvalues and eigenfunctions are considered as *frequencies* $(2\pi\lambda)$ and the *power* $(K(2\pi\lambda))$; details of equations and algorithms can be found [15]. Hence, FI_K is defined as follows,

$$FI_K = \frac{\sum_{\lambda=H_1+1}^{H_2} K(2\pi\lambda)}{\sum_{\lambda=L+1}^{H_1} K(2\pi\lambda)} \qquad (4.8)$$

where $H_1 = \frac{3N_{FFT}}{f_s}$, $H_2 = \frac{8N_{FFT}}{f_s}$, $L = \frac{0.5N_{FFT}}{f_s}$, f_s is sampling frequency.

4.4.3 Anomaly Scores

Basically, the anomaly score of a feature value at time n, $A(n)$, determines if the feature value extracted from a window at time n is higher than a threshold (i.e., is an anomaly). In this case study report, $A(n)$ is calculated using Eq. (3.5) (Sect. 3.2.1) with the scale factor $\alpha = 1$ (i.e., a simple case of no scaling deviation).

4.4.4 Exploratory Pool

We construct a feature pool that consists of 244 features (Table 4.1). The first half of the pool are 122 candidates, extracted using seven previously published features and our four aforementioned new features. Existing extraction methods include average, standard deviation, variance, median, entropy, energy, power and *FI* as found [2, 5, 9, 13, 20, 21, 23, 24, 38]. New methods consist of the maximum and number of peaks of C_{XY} in the spectral coherence [4], *multi-channel FI* (FI_{MC}), and the Koopman spectral analysis (FI_K) [17]. These eleven extraction functions are applied to single and multiple inputs. Specifically, FI_{MC} and FI_K are applied to MCMS while the other functions are to SCSSs and the sum square of all three channels of single sensors. The second half of the pool consists of 122 anomaly score vectors (Sect. 4.4.3) of the above 122 features.

Table 4.1 Feature pool for FoG examined in this work. Three types of data inputs: SCSS, MCSS, MCMS. Identifications of sensors and axes describe for each channel used to extract features. Eight existing functions and two new of C_{XY} are for SCSS and MCSS. For MCMS, two new features are FI_K (by Koopman spectral analysis) and FI_{MC} (by Fourier transform Eq. (4.5). Feature IDs 123 $\rightarrow$ 244 are corresponding anomaly score based features of the above 1 $\rightarrow$ 122

Sensor	SCSS									MCSS			MCMS	
	0 (ankle)			1 (knee)			2 (back)			0	1	2	0,1,2	0,1,2
Axis	x	y	z	x	y	z	x	y	z		$\sqrt{x^2+y^2+z^2}$		x,y,z	x,y,z
Extraction	Average, standard deviation, variance, median, entropy, energy, power, FI, C_{XYNpks}, and C_{XYmax}												FI_K	FI_{MC}
IDs	1:10	11:20	21:30	31:40	41:50	51:60	61:70	71:80	81:90	91:100	101:110	111:120	121	122

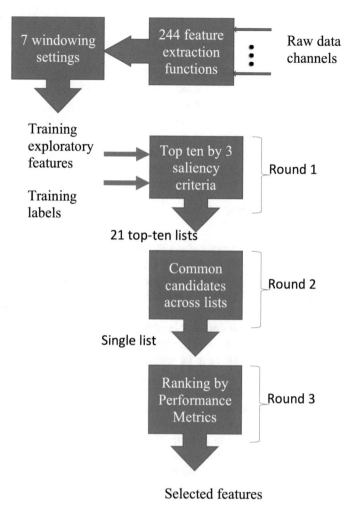

Fig. 4.5 Feature selection process. 244 features as described in Table 4.1. 7 window sizes are $2 \rightarrow 8$ s in steps of 1 s. Three saliency criteria are DIS, MI, Var-Ratio scores. Common candidates are entries that are shared by more than one list of *Round1*

4.4.5   *Feature Selection*

We propose a voting process to select the best feature from the large exploratory pool (as introduced in Sect. 2.3). This process uses three levels of selection: saliency, robustness, and accuracy; called *Round1, Round2, Round3* respectively (Fig. 4.5). After each level, selected candidates become more favourable. Specifically, *Round1* suggests the most salient and discriminative subset. Then, *Round2* examines if the candidates are robust across window sizes. Finally, *Round3* tests the detection performance of these features using our ASD.

In *Round1*, feature candidates are ranked according to three saliency criteria, i.e., mutual information (MI), separability calculated using Euclidean distances (DIS), and the variance ratio of clusters (Var-Ratio). This step is implemented across 7 window sizes ($2 \rightarrow 8\,s$ in steps of $1\,s$), creating 21 lists of ranking scores. The range for window sizes is based on the minimum and maximum values currently suggested in the literature (e.g., $2\,s$ [38] and $7.5\,s$ [24]). After finding a subgroup of high saliency score, the robustness is examined in *Round2*. Secondly, salient candidates are identified that shared in more than one list across window sizes or criteria (i.e., robustness). Finally, accuracy metrics are used to find the subset for our ASD.

4.5   Performance Metrics

In the literature, automatic techniques have been evaluated using different measures such as confusion matrices and/or intra-class correlations (ICCs) [36]. For instance, authors of [2, 20, 21] used timing-instance-based confusion matrices (i.e., counting FoG time frames and often involving a tolerance of milliseconds or seconds); and authors of [9, 23, 24, 38] used event-based confusion matrices (i.e., counting continuous FoG epochs) and ICCs on the number of FoG events or percentage of freezing time over a trial. With regard to real-time applications using wearable FoG detectors, the timing-based method is of most interest, whereas event-based is important in clinical FoG assessments. We utilize both types during feature selection as extra criteria (apart from saliency scores).

In our work, ICCs are used as supplemental criteria during *Round3* to select features rather than in performance comparisons with other works due to several limitations of ICC usages; e.g., intra-rater reliability reported for FoG number was only 0.44 and at least two observers are recommended to analyse task videos [26]. In this work, information regarding the reliability for manual ratings were not available (nor were the number of raters). Thirdly, walking tasks were designed to have a single recording session per subject (about 30 min) rather than several short trial recordings (around one minute each). Hence, because in our data set the number of individual recordings is relatively small, thus, data are grouped into one-minute segments. We assume that the segmentation is close to the multi-trials settings. Therefore, our estimation of ICC is a non-decreasing relationship with the reported ICC in the literature. Given two vectors of an automatic detection result and manual labels, the estimated intraclass correlation is calculated as [22]; specifically the ICC(A-1) designation is used (two-way random effects) for the degree of absolute agreement among measurements.

With respect to the timing-based metrics, in confusion matrices, *ground truth* is referred to the manual video analysis, and *positives* for FoG windows. True Positives (TP) are windows which were marked as FoG by both a test algorithm and the label. False Positives (FP) are windows labelled as FoG but did not agree with the *ground truth*. Windows that the test method failed to label as FoG but were annotated as such, are defined as False Negatives (FN). When the test method and the human agree a window was non-FoG, it is counted as a True Negative (TN). Please note that the reference labels used in this work were made by human thus are subjective. Likewise the literature works [2, 20], we investigate a tolerance, *tol*. Let t be the time instance an automated method decides it is FoG. If within the range of $[t - tol, t + tol]$, there is at least one instance where the reference (i.e., manual method) says it is FoG, we count this agreement is a true positive. Otherwise it is a false positive. Similarly for negative cases. The tolerance will be determined during the experiments using the performance curves (ROC).

Sensitivity and specificity are $\frac{TP}{TP+FN}$ and $\frac{TN}{TN+FP}$, respectively. F1-score, which is the harmonic mean of precision and sensitivity, with best value at 1 and worst at 0 [30], is calculated as $\frac{2TP}{(2TP+FP+FN)}$.

4.6 Results

4.6.1 Selection by Saliency

Three types of ranking scores (i.e., MI, DIS, and Var-Ratio) across window sizes for each feature candidate were measured (Fig. 4.6). An example feature ranking with window size of $2s$ using three saliency metrics are illustrated in Fig. 4.6 (a, b, c) (sorted from high to low scores). The order of ranking is from 1 to 244 (high to low); a higher saliency score indicates the higher ranking order. The other window sizes shared a similar trend.

As can be seen, scores outside the top ten rank (outside of the dotted vertical line) dropped quickly. Therefore, these ten candidates were selected for further steps. Specifically, *Round1* contains 210 entries of 21 short lists. We noticed that there were only 64 distinct features in *Round1*. For example, Fig. 4.6d to k illustrate that the top-ten lists share many features. In these sub-plots, new features are indicated with circle markers and labelled horizontal axes with feature identifications (IDs). Description of IDs can be found in Table 4.1. Our shortlists include $FI0y$ (i.e., freezing index from ankle at vertical axis [2, 20, 23]) and previously proposed features (e.g., $FI2y$ [24], $FI2x$ [38], energy, sum of power $Psum$ [2], and their standard deviation, mean, variance [21]). Among the 64 distinct features, our new candidates, C_{XYNpks}, C_{XYmax}, FI_K, and FI_{MC}, were listed in more top-ranking lists than the existing ones.

4.6.2 Selection By Robustness

In the second round, members of *Round1* that are selected as the top ten in more than one list (across window sizes and/or criteria) are considered robust features. There were 33 entries in *Round2* (i.e., about half of *Round1*). Interestingly, FI_{MC} is one of the most robust candidates in terms of being selective across window sizes

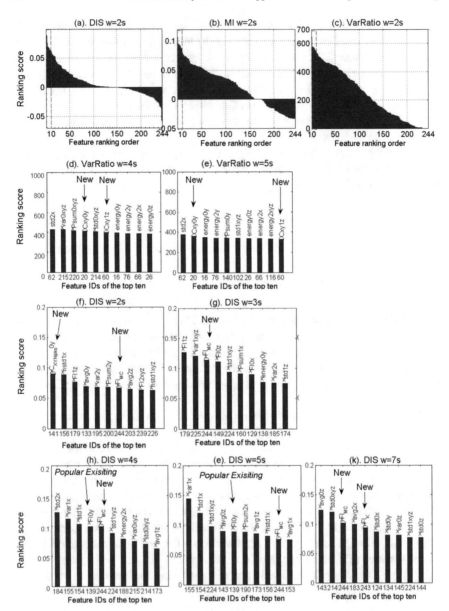

Fig. 4.6 Example of feature ranking and the shortlists. **a**, **b**, **c** are ranking scores for the feature pool; **a** DIS, **b** MI, **c** Var-Ratio scores. Vertical axes: saliency scores. Horizontal axes: ranking order. The top-ten lists are in the dotted boxes. The others, **d–k**, illustrate the sharing among shortlists across window sizes and criteria. Features with circle markers are new while others are have been currently used in literature. The top ten identifications (IDs) of features are detailed in Table 4.1

(Table 4.2). Other new or popular existing candidates are also added in the table for comparison purposes.

4.6.3   Selection By Detection Performance

In the third round, a simple form of ASD (Sect. 3.2.1) is used to rank features in *Round2* by performance criteria. We also examined our other new or popular existing features (Table 4.2) for comparison purposes. ICCs results for freezing time percentage and number of FoG showed only seven candidates that had at least one report of $ICC > 0.2$ (suggestion from [26]), as shown in Fig. 4.7. These candidates are FI_{MC}, FI2y, FI2x, FI1z, FI0y, Mean 0z, and Mean 1z (Table 4.1).

During training period, the receiver operating characteristic (ROC) is calculated for each window size of each feature extraction with a timing tolerance range from $0 \rightarrow 1$ *s* in steps of 0.1 *s*. We observed that configurations FI_{MC} (3 *s*), FI0y (2 *s* or 7 *s*), FI1z (6 *s*), FI2y (3 *s* or 8 *s*), called *Round3*, had excellent results (Fig. 4.8). Due to the difficulty of visualizing ROCs across many variables, F1-scores were displayed in Fig. 4.8 instead.

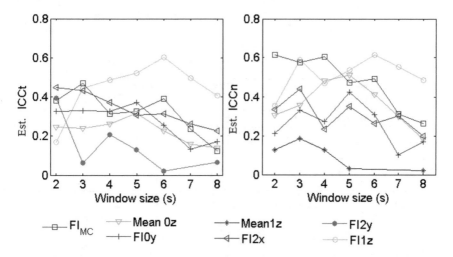

Fig. 4.7 ICCs for feature selection in FoG detection. Markers are for different features. Estimated ICC for the freezing time percentage (Left) and number of FoG events (Right)

Table 4.2 Thirty-three top salient and robust features (*Round2*) and four others of interest. IDs are identifications of features. 'Std': standard deviation. DIS, MI, and Var-Ratio are criteria

Feature ID	Name	Sensor, Channel	Window sizes (second)		
			DIS	MI	Var-Ratio
244	FI_{MC}	all	all	–	–
194	Std	2,y	–	all	2
124	Std	0,x	8	all	3
214	Std	0,xyz	4,8	all	4,6,7,8
134	Std	0,y	8	all	6,8
154	Std	1,x	4,5	all	–
174	Std	1,z	3	all	–
184	Std	2,x	4, 5, 6	2	all
138	Energy	0,y	3	–	all
98	Psum	0xyz	–	3 → 7	–
224	Std	1xyz	3 → 8	–	–
155	Variance	1x	4 → 7	–	–
198	Energy	2y	6	–	2 → 5,7
26	Energy	0z	–	–	2 → 5,7
102	Std	1xyz	–	2 → 5	5,7
164	Std	1y	7	2 → 5,8	–
215	Variance	0xyz	–	–	4 → 8
188	Energy	2x	4	–	2 → 5
112	Std	2xyz	6	2 → 5	–
175	Variance	1z	7	6,7,8	3,6
158	Energy	1x	7	–	6,7,8
22	Std	0z	7,8	6,7,8	–
116	Energy	2xyz	–	–	2,3,5
135	Variance	0y	–	–	6,8
93	Variance	0xyz	–	6,7,8	–
185	Variance	2x	3,6	–	–
139	*FI* [23]	0y	4,5	–	–
173	Mean	1z	4,5	–	–
179	FI	1z	2,3	–	–
225	Variance	1xyz	3,7	–	–
143	Mean	0z	5,8	–	–
20	*Cxymax*	0y	–	–	4,5
60	*Cxymax*	1z	–	–	4,5
195	Variance	2y	2,6	–	2
Other new or existing features for comparison purposes					
141	*CxyNpks*	0y	2	–	–
243	FI_K	all	8	–	–
199	FI [24]	2y	6	–	–
189	FI [38]	2x	–	–	–

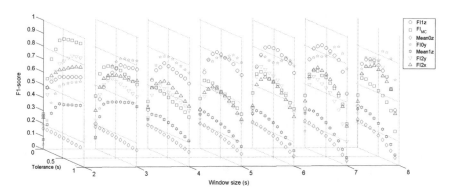

Fig. 4.8 Effects of window sizes and tolerances on F1-scores of ASD. Tolerance from $0 \to 1\ s$. Three dimensional view for windows from $2 \to 8\ s$. Markers are for different features

4.6.4 Tests and Comparisons with the Same Cohort Set

We then applied unseen test sets (five subjects who have been with PD for 11.2 ± 9.6 years; H&Y score: 2.9 ± 0.74) to validate ASD. Three subjects had FoG during data collection while the other two had no FoG. We noticed that, during the validation, FI_{MC} (3 s) and FI2y (3 s) had high accuracies with lowest deviation between training and out-of-sample tests (Table 4.3). FI0y, a popular feature in existing detectors (7 s windows), achieved a sensitivity of 79% (specificity of 79.5%) at a tolerance of 0.3 s. On the other hand, FI0y scores the highest F1-score of 84% with 2s window size (tolerance of 0.9 s). FI2y with 8 s windows and 0.9 s tolerance can achieve sensitivity of 87.5% (specificity of 84.5%).

Hence, we propose an optimization configuration for ASD as follows: window size as small as 3 s, tolerance for performance measurements of 0.4 s, freezing index is used for feature extraction. There was a slight preference of sensor locations between ankle and hip in terms of further performance improvement.

Table 4.3 Development performance of ASD using features in *Round3*. 'Win': window size. 'Tol': tolerance. 'SD': standard deviation of development and out-of-sample test. Performance in %. Sens: Sensitivity. Spec: Specificity. F1: F1-score

Feature ID	Name Channel	Parameter		Development (%)		Out-of-sample (%)		Average ± SD (%)		
		Win.	Tol.	Sens	Spec	Sens	Spec	Sens	Spec	F1
244	FI_{MC}	3s	0.2s	85	74.0	77.0	80.0	81.0 ± 6	77.0 ± 4	74.5 ± 6
139	FI 0y	2s	0.9s	88.0	81.0	86.0	63.0	87.0 ± 1	72.0 ± 13	84.0 ± 10
		7s	0.3s	71.0	93.0	87.0	66.0	79.0 ± 11	79.5 ± 19	82.5 ± 11
179	FI 1z	6s	0.1s	80.3	80.0	82.0	58.0	81.0 ± 1	69.0 ± 16	78.0 ± 6
199	FI2y	3s	0.4s	75.0	80.0	83.0	92.0	79.0 ± 6	86.0 ± 8	76.5 ± 12
		8s	0.9s	76.0	74.0	99.0	95.0	87.5 ± 16	84.5 ± 15	82.0 ± 24

4.6.5 External Validation Tests

Finally, using independent test sets that were from a different cohort to the one we used for training (Sect. 4.3.2), we validated our proposed ASD-based method (i.e., online ASD detector, freezing index feature, window size of 3 s). Though the performance improvement between ankle and hip sensor locations was not significant during the development stage, for a better comparison with existing works that used both types of inputs: single channel and multiple channels, such cases were still included in our report. Table 4.4 shows its high accuracy performance comparing with earlier works across several configurations of inputs.

Table 4.4 Out-of-sample detection performance of ASD (versus existing methods [2][a] [20][b] [24, 38][c,d]) across configurations[e] and datasets[f]. Performance in %

Method	Settings			Performance (%)		
	Input	*Win*	*Tol*	Sensitivity	Specificity	F1
CNR [2]	FI0y, Psum0y	4s	2s	73.1[a]	81.6[a]	–
Learning [20]	Mean0y, Std0y, FI0y, Energy0y	4s	1s	66.25[b]	95.38[b]	–
Global [24]	FI012y[d], $\overline{FI} = 3$	7.5s	–	84.3[c]	78.4[c]	–
Global [38]	FI2x, $\overline{FI} = 1.47$	2s	–	75.0[c]	76.0[c]	–
Online ASDs (proposed), external validatio[f]						
ASD multi-inputs	FI_{MC}	3s	0.4s	96 ±17	79 ±41	99 ±7
ASD ankle y-axis	FI0y	3s	0.4s	94 ±23	84 ±36	99 ±4
ASD hip y-axis	FI2y	3s	0.4s	89 ±32	82 ±39	96 ±18
ASD hip x-axis	FI2x	3s	0.4s	89 ±32	94 ±23	97 ±17

[a] as reported [2] using CNR classifier and LOOCV
[b] as reported [20] using Random Forest classifier and LOOCV
[c] for event-based calculation while others were for timing-based
[d] the majority vote of *seven sensors* [24].
[e] *Input*: features, sensors, and axes. '*Tol*': tolerance. '*Win*': window size
[f] 71 trials of 15 subjects; different cohort to the training set (same to the work [24])

4.7 Discussion

During the development stage, we observed that beside the existing FI extracted from ankle sensor at vertical axis, our new feature with multiple channels, FI_{MC}, is one of the top features in saliency, clusterability, and robustness. Only seven out of 244 candidates met requirements of our three-round selection procedure. To detect FoG, we implemented an anomaly score based detector, ASD. With ASD, our features outperformed existing works with a small window and/or low tolerance. Specifically, FI2y, the freezing index from vertical data at a hip sensor, was found to be the best choice for performance; achieving sensitivity (specificity) of 87.5% (84.5%) with 8 s windows and 0.9 s tolerance. FI_{MC}, is also a promising candidate. For example, FI_{MC} has high ICC and is the most robust candidate across window sizes during feature selection by saliency. FI_{MC} achieved a sensitivity of 81% (specificity of 77%) at the smallest tolerance of 0.2 s (3 s windows).

During the test stage, we reported out-of-sample test outcomes in as many similar configurations as suggested from compared works. Our ASD that performed better than current methods can use only one type of feature extraction (freezing index) from a single channel. It is flexible and convenient to choose a sensor location between ankle and hip. Our proposed method significantly outperforms (e.g., mean ($\pm$SD) of sensitivity, specificity are 94% ($\pm$23%) and 84% ($\pm$36%) for *ASD ankle y-axis*) other automated methods in the literature.

Regarding the system design, to the best of our knowledge, [5, 20] achieved the best published performance to date for subject-independent settings. Specifically, with different reported configurations, these two methods used a *context recognition network* [2] and a Random Forest [20] with leave one out cross validation techniques (LOOCV). Other works used various *global $\overline{FI}$* values with different channel selection. Note that, sensitivities and specificities [23, 24, 38] were for event-based calculation that may differ from the others in Table 4.4. Our detector is an anomaly detection technique that has low computational cost and is feasible for real-time operation in a subject-independent manner. As presented, our performance is significantly higher than the one of compared automatic detectors while using a much smaller window and/or lower tolerance.

4.8 Summary

In this chapter, one successful application of our novel supervised voting technique for feature engineering with application to point anomaly detection in FoG monitoring is demonstrated. Previous methods have utilised features yielding high variability across time and subjects. The new features found using our technique are not only more sensitive, they also have lower temporal and subject-dependent variation. These features were exploited to achieve an improve anomaly detector with low computational cost.

References

1. Ackerman M, Ben-David S (2009) Clusterability: a theoretical study. In: International conference on artificial intelligence and statistics, pp 1–8
2. Bachlin M, Plotnik M, Roggen D, Maidan I, Hausdorff J, Giladi N, Troster G (2010) Wearable assistant for Parkinson's disease patients with the freezing of gait symptom. IEEE Trans Inf Technol Biomed 14(2):436–446
3. Bloem B, Hausdorff J, Visser J, Giladi N (2004) Falls and freezing of gait in Parkinson's disease: a review of two interconnected, episodic phenomena. Mov Disord 19(8):871–884
4. Challis R, Kitney R (1990) Biomedical signal processing (part 3 of 4):the power spectrum and coherence function. Med Biol Eng Comput 28(6):509–524
5. Cole B, Roy S, Nawab S (2011) Detecting freezing-of-gait during unscripted and unconstrained activity. In: Annual international conference of the IEEE engineering in medicine and biology society, EMBC, pp 5649–5652
6. Daubechies I, Bates BJ (1993) Ten lectures on wavelets. J Acoust Soc Am 93(3):1671–1671
7. Fahn S, Elton R (1987) Unified rating scale for Parkinson's disease. In: Recent developments in Parkinson's disease, pp 153–163
8. Folstein MF, Folstein SE, McHugh PR (1975) mini-mental state: a practical method for grading the cognitive state of patients for the clinician. J Psychiatr Res 12(3):189–198
9. Gazit E, Bernad-Elazari H, Moore S, Cho C, Kubota K, Vincent L, Cohen S, Reitblat L, Fixler N, Mirelman A et al (2015) Assessment of Parkinsonian motor symptoms using a continuously worn smartwatch: preliminary experience. Mov Disord 30:S272–S272
10. Gibb WRG, Lees A (1988) A comparison of clinical and pathological features of young-and old-onset Parkinson's disease. Neurology 38(9):1402–1402
11. Giladi N, Tal J, Azulay T, Rascol O, Brooks DJ, Melamed E, Oertel W, Poewe WH, Stocchi F, Tolosa E (2009) Validation of the freezing of gait questionnaire in patients with Parkinson's disease. Mov Disord 24(5):655–661
12. Goetz CG, Tilley BC, Shaftman SR, Stebbins GT, Fahn S, Martinez-Martin P, Poewe W, Sampaio C, Stern MB, Dodel R et al (2008) Movement Disorder Society-sponsored revision of the Unified Parkinson's Disease Rating Scale (MDS-UPDRS): scale presentation and clinimetric testing results. Mov Disord 23(15):2129–2170
13. Han J, Lee W, Ahn T, Jeon B, Park KS (2003) Gait analysis for freezing detection in patients with movement disorder using three dimensional acceleration system. In: Proceedings of the 25th annual international conference of the IEEE engineering in medicine and biology society, vol 2, pp 1863–1865
14. Hoehn MM, Yahr MD (1998) Parkinsonism: onset, progression, and mortality. Neurology 50(2):318–318
15. Hua JC, Roy S, McCauley JL, Gunaratne GH (2016) Using dynamic mode decomposition to extract cyclic behavior in the stock market. Phys A Stat Mech Appl 448:172–180
16. Kira K, Rendell LA (1992) The feature selection problem: traditional methods and a new algorithm. In: Proceedings of the tenth national conference on artificial intelligence, AAAI Press, AAAI'92, pp 129–134. http://dl.acm.org/citation.cfm?id=1867135.1867155
17. Koopman BO (1931) Hamiltonian systems and transformation in Hilbert space. Proc Natl Acad Sci USA 17(5):315
18. Latt M, Lord S, Morris J, Fung V (2009) Clinical and physiological assessments for elucidating falls risk in Parkinson's disease. Mov Disord 24(9):1280–1289
19. Macht M, Kaussner Y, Moller J, Stiasny-Kolster K, Eggert K, Kruger H, Ellgring H (2007) Predictors of freezing in Parkinson's disease: a survey of 6,620 patients. Mov Disord 22(7):953–956
20. Mazilu S, Hardegger M, Zhu Z, Roggen D, Troster G, Plotnik M, Hausdorff J (2012) Online detection of freezing of gait with smartphones and machine learning techniques. In: 6th international conference on pervasive computing technologies for healthcare (PervasiveHealth), pp 123–130

21. Mazilu S, Calatroni A, Gazit E, Roggen D, Hausdorff JM, Tröster G (2013) Feature learning for detection and prediction of freezing of gait in Parkinson's disease. In: Machine learning and data mining in pattern recognition. Springer, pp 144–158

22. McGraw KO, Wong SP (1996) Forming inferences about some intraclass correlation coefficients. Psychol Methods 1(1):30

23. Moore S, MacDougall H, Ondo W (2008) Ambulatory monitoring of freezing of gait in Parkinson's disease. J Neurosci Methods 167(2):340–348

24. Moore ST, Yungher DA, Morris TR, Dilda V, MacDougall HG, Shine JM, Naismith SL, Lewis SJG (2013) Autonomous identification of freezing of gait in Parkinson's disease from lower-body segmental accelerometry. J Neuroeng Rehabil 10(1):1

25. Moreau C, Defebvre L, Bleuse S, Blatt J, Duhamel A, Bloem B, Destée A, Krystkowiak P (2008) Externally provoked freezing of gait in open runways in advanced Parkinsons disease results from motor and mental collapse. J Neural Transm 115(10):1431–1436

26. Morris TR, Cho C, Dilda V, Shine JM, Naismith SL, Lewis SJ, Moore ST (2012) A comparison of clinical and objective measures of freezing of gait in Parkinson's disease. Parkinsonism Related Disord 18(5):572–577

27. Paul S, Canning C, Sherrington C, Lord S, Close J, Fung V (2013) Three simple clinical tests to accurately predict falls in people with Parkinson's disease. Mov Disord 28(5):655–662

28. Pham TT, Moore ST, Lewis SJG, Nguyen DN, Dutkiewicz E, Fuglevand AJ, McEwan AL, Leong PH (2017) Freezing of gait detection in Parkinson's disease: a subject-independent detector using anomaly scores. IEEE Trans Biomed Eng 64(11):2719–2728

29. Reimer J, Grabowski M, Lindvall O, Hagell P (2004) Use and interpretation of on-off diaries in Parkinson's disease. J Neurol Neurosurg Psychiatry 75(3):396–400

30. Rijsbergen CJV (1979) Information retrieval, 2nd edn. Butterworth-Heinemann, Newton, MA, USA

31. Schaafsma J, Balash Y, Gurevich T, Bartels A, Hausdorff J, Giladi N (2003) Characterization of freezing of gait subtypes and the response of each to levodopa in Parkinson's disease. Eur J Neurol 10(4):391–398

32. Schmid PJ (2010) Dynamic mode decomposition of numerical and experimental data. J Fluid Mech 656:5–28

33. Sejdi E, Lowry KA, Bellanca J, Redfern MS, Brach JS (2014) A comprehensive assessment of gait accelerometry signals in time, frequency and time-frequency domains. IEEE Trans Neural Syst Rehabil Eng 22(3):603–612

34. Shannon C (1948) A mathematical theory of communication. Bell Syst Techn J 27(3):379–423

35. Shine J, Moore S, Bolitho S, Morris T, Dilda V, Naismith S, Lewis S (2012) Assessing the utility of freezing of gait questionnaires in Parkinsons disease. Parkinsonism Relat Disord 18(1):25–29

36. Shrout PE, Fleiss JL (1979) Intraclass correlations: uses in assessing rater reliability. Psychol Bull 86(2):420

37. Snijders AH, Weerdesteyn V, Hagen YJ, Duysens J, Giladi N, Bloem BR (2010) Obstacle avoidance to elicit freezing of gait during treadmill walking. Mov Disord 25(1):57–63

38. Zach H, Janssen AM, Snijders AH, Delval A, Ferraye MU, Auff E, Weerdesteyn V, Bloem BR, Nonnekes J (2015) Identifying freezing of gait in Parkinson's disease during freezing provoking tasks using waist-mounted accelerometry. Parkinsonism Relat Disord 21(11):1362–1366

Chapter 5
Collective Anomaly Detection: Application to Respiratory Artefact Removals

5.1 Background on Respiratory Artefact Removal in FOT Data

The forced oscillation technique (FOT) [11] is a lung function test that can provide useful information from short duration recordings, and only requires passive cooperation from the subject [30]. FOT assesses breathing mechanics by superimposing small external pressure signals to the spontaneous breathing of the subject. A total respiratory mechanical impedance (Zrs), which includes airway resistance together with elastic and inertive behavior of the lungs and the chest wall, is then measured at one oscillation frequency (mono-frequency oscillations) or several (multi-frequency). Zrs is described as a complex number with *real* and *imaginary* components, called the resistance (Rrs) and reactance (Xrs) respectively. A primary reason hindering its widespread adoption lies in difficulties associated with removing artefacts. This results in lower reproducibility than the most common pulmonary function test, or spirometry. Manual removal by operators, called the *human-based* method, is currently considered the gold standard for respiratory artefact removal practice. This, however, is typically done in an ad-hoc manner which is laborious, and subjective.

To detect artefacts, several automated refinements include detecting low (e.g., transducer noise) and high frequency artefacts (e.g., light coughing, mouth piece leak, swallowing, glottic closure and tongue occlusion) are necessary. According to the quality control guidelines [17], low frequency noise removal rejects low magnitude-squared coherence values of pressure and flow [26]. Several transient artefacts are removed by identifying deviations from the norm, called *thresholding* approaches.

To exclude respiratory artefacts, two different strategies are point rejection [5, 6, 28] and complete-breath rejection [24, 26]. For example, a point-based method called *3SD* [28], introduced a statistical filter that rejected any impedance points greater than three standard deviations (SD) from the mean Rrs or Xrs value. Alternatively, the *complete-breath* approach rejects entire breaths as defined by the starting and ending points of breath cycles in which at least one data point is out of the *3SD*

© Springer Nature Switzerland AG 2019
T. T. Pham, *Applying Machine Learning for Automated Classification of Biomedical Data in Subject-Independent Settings*, Springer Theses, https://doi.org/10.1007/978-3-319-98675-3_5

range [26], called *B-3SD* method. The complete-breath rejection has been reported to be more accurate than the point approach as it can avoid an imbalance between the inspiratory and expiratory contributions to each breath [26]. Nonetheless, these automated attempts still miss numerous artefacts.

5.2 Data Collection

5.2.1 Subjects and Protocol

We collated data from two different age groups (*Paediatrics* and *Adults*, Table 5.1). The paediatric dataset comprised a random sample of 9 subjects (total 69 FOT runs) for training and 5 subjects (total 31 runs) for out-of-sample tests. These were taken from a much larger ongoing epidemiological study, which has been described in detail elsewhere (Ultrafine Particles from Traffic Emissions and Children's Health, UPTECH) [13, 19]. The epidemiological study collected FOT data, as part of its respiratory function assessment, in eight- to eleven-year-old children recruited from 25 different public primary schools in Queensland, Australia. FOT was performed for at least 30 min after supervised medication administration and with at least 10 min rest prior to recording. Zrs was measured at 6 Hz, using an in-house built FOT device (transducer Sursense DCAL-4, Honeywell Sensing and Control; more details are available [32]) and modification to comply with recent recommendations [4]. Children were encouraged to breathe in a regular manner, avoid swallowing and maintain a tight mouthpiece seal. Children had multiple recordings in a single session as part of the study protocol.

For the adult group, 9 healthy participants and 10 asthmatic patients were recruited from staff and patients of the Royal North Shore Hospital, St Leonards, Australia and the Woolcock Institute of Medical Research volunteer database (Glebe, Australia) [34]. Healthy participants were non-smokers with no known respiratory disease. Asthmatic adults had a physician diagnosis of asthma (clinically stable as defined by GINA guidelines [3]) and had no reported diagnoses of any other cardiac or pulmonary disease. The asthmatic and control subjects had three recordings over seven days within a 10-day period at the Respiratory Investigation Unit at Royal North Shore Hospital [34]. To ensure clinical stability, asthmatic patients continued to take their usual medications and were reviewed by a specialist physician at each visit for any changes in their usual symptoms. All recordings were performed at the same schedule to avoid any diurnal variation effects. Zrs was measured at an oscillation frequency of 6 Hz from a FOT device of similar general design and specifications as the children dataset [21]. Three separate consecutive recordings were collected with subjects breathing tidally for 60 s at each session (day). The participants put their nose clip on and placed their hands on cheeks to reduce the upper airway shunt. Recordings were assessed from visual inspection by a technician if tidal volume and breathing frequency appeared stable. Artefact labels were made by the operator

Table 5.1 Descriptions of FOT data sets used in this work

Dataset	Subjects	Recordings	Breaths	Description
Ds1	9	69	1110	Development, children (asthmatics, Westmead Hospital)
Ds2	9	261	3067	Development, adults (healthy, Woolcock Institute)
Ds3	5	31	580	Test, children (asthmatics, Westmead Hospital)
Ds4	10	285	3947	Test, adults (asthmatics, Woolcock Institute)

Table 5.2 Subject characteristics of development and test sets

Characteristics	Children	Adults
N (Subjects)	15	20
Measurements (Recordings)	70	546
Total breaths	1690	7014
Mean (±SD) age, years	10.4 (± 1.1)	Healthy = 32.2 (±5.9); Asthma = 37.5 (±11.6)
Other	Weight (kg) = 33.56 (±6.73)	Body massindex: Healthy = 23.2 (±1.5)
	Height (cm) = 137.42 (±6.47)	Asthma = 25.2 (±4.6)

using recommendations [26] (more details [34]). All subjects gave written, informed consent and the study was approved by The Human Research Ethics Committee of Northern Sydney Central Coast Health (protocol no. 0903-050M). For children, the study was approved by the Queensland University of Technology Human Research Ethics Committee (Table 5.2).

5.2.2 Data Pre-processing

Flow was measured using a screen type pneumotachograph (3100 series, flow range 0–160 L/min). Flow and pressure signals were digitally sampled at 396 Hz and bandpass filtered with a bandwidth of ±2 Hz centred around 6 Hz. Breath cycles were defined as described in the preliminary work [13, 22, 23, 34].

Rrs and *Xrs* were calculated at 0.1 s intervals using a standard frequency-domain method. To ensure balance between the inspiratory and expiratory contributions to each breath [26], incomplete or partial breaths at the beginning/end of the recording were removed. Since the "not accepted" annotations included non-eligible physiological breaths which are commonly known to be rejected by the standard FOT quality guidelines [17], we discard these artefacts in pre-processing steps and report separately in later comparisons. First, we remove breaths that contain negative *Rrs* which are non-physiological. Then, we discard breaths that have magnitude-squared coherence values of pressure and flow less than 0.9 [26]. Unusually high amplitude observations were successfully caught by the *B-3SD* approach [26] and discarded. Finally, we apply *3IQR* (i.e., *3IQR* away from the median) to *Rrs*, *Xrs*, *Volume*, *Pressure*, and *Flow*.

5.3 Performance Metrics

True Positives (TP) are breaths which were marked as "artefacts" by both a test algorithm and the annotation. False Positives (FP) are breaths we labeled as artefacts but did not agree with the ground truth. Breaths that we failed to label as artefacts but were annotated as such, are defined as False Negatives (FN). When the test method and the human agree that a breath was not an artefact, it is counted as a True Negative (TN).

Sensitivity and specificity are $\frac{TP}{TP+FN}$ and $\frac{TN}{TN+FP}$, respectively. F1-score, which is the harmonic mean of precision and sensitivity, has best value at 1 and worst at 0 [25], is calculated as $\frac{2TP}{(2TP+FP+FN)}$. *Throughput* is the ratio of breath numbers in the output to input, $\frac{TN+FN}{total\ input}$. *Approval rate* of the filtered data (i.e., the breaths remaining after removal) is the ratio of breaths that are "accepted" by the human to the total output breaths, $\frac{TN}{TN+FN}$.

Since *Rrs* is one of the main outcomes of FOT in clinical and research usage, we consider *variability* (i.e., the standard deviation divided by mean) of the average *Rrs* for each patient to be a critical metric. Specifically, within-session coefficients of variation (wCV) and/or between-session (bCV) of measurements for average values of *Rrs* recordings within one day of recording (i.e., wCV) or across days (i.e., bCV).

To quantify this, we aim for an equivalent average *Rrs*, and lower or equal SD compared with the human-based approach. However, if we only consider variability, we may not account for the number of valid breaths that remain, e.g., we may discard most valid breaths together with invalid ones to achieve low variability. Therefore, when comparing techniques, we should strive for an equivalent preservation level *and* lower variability.

The *preservation* after removal can be summarised by standard accuracy metrics (e.g., sensitivity, specificity, and F1-score [25]) and our new metrics: *throughput* and *approval rate*. In confusion matrices for accuracy calculation, we consider *groundtruth* to be the human labels (or "manual"), and *positives* to be artefacts.

5.4 Proposed Artefact Detection Scheme

From a machine learning perspective, each breath is represented by a vector of features. Features are then classified by a model (*detector*) constructed from domain-knowledge and/or human annotations (labels). The aforementioned existing automated methods are unsupervised techniques in which plain feature extraction is often exploited and threshold values are chosen as a number of standard deviations away from the mean of a single measurement. We hypothesize that advanced extraction (e.g., two-dimensional, 2D) may provide more relevant features in order to alleviate the above limitation of current existing automated methods. The relevance of novel features can be confirmed non-heuristically by supervised techniques (*feature selection*). Specifically, correlation of feature candidates with artefact characteristics can be measured by mutual information (Shannon's information theory [29]). The clusterability of a candidate [1] indicates the efficiency of using threshold values to detect artefacts. Two typical ways to assess the clusterability are the variance ratio of clusters [1] and the separability calculated by Euclidean distances from an instance to a *near-hit* and *near-miss* [15].

Given an exploratory feature pool, by selecting the k highest ranking candidates (e.g., $k = 10$ often used in literature of feature selection), we can construct a more accurate anomaly detector as non-salient features which cause overfitting are discarded. Several challenging factors should be noted. One is the time-dependency of lung function (e.g., lung elasticity [20]). The others are clinical aspects of FOT (e.g., *Rrs* and *Xrs* are dependent on body size and possibly racial/ethnic differences [30]). Thus, to avoid dependency, feature ranking scores should be accumulated across recordings. We also noticed that *Rrs* within a recording can be non-Gaussian, with a strong kurtosis. Hence, when applying threshold values, we do not assume a particular distribution. Instead we use quartile percentages (called *quartile thresholding*). In contrast to earlier works, the deviation threshold is also not assumed, rather it is determined from the receiver operating characteristic (ROC) and other performance metrics with training datasets. In this work, we evaluate FOT measurements at single frequency and expect similar observations with other frequencies.

5.5 Feature Extraction

After data preprocessing, we construct a pool of exploratory features. According to a ranking report of these candidates, we select the most salient subset of features (Sect. 3.1) for further detection steps. The pool consists of 111 candidates (Table 5.3), of which 11 have been previously reported. Our new features include *landmark* information and *resampling* values.

Table 5.3 List of FOT data features examined in this work

Measurement	Domain	Function description	New?	ID
Pressure	Frequency	Maximum value of first level DWT	No	1
Xrs	Time	Maximum, minimum, range	No	2–4
Xrs	Time	20-point resampled	Yes	8–27
Volume	Time	Maximum, minimum, range	No	5–7
Volume	Time	20-point resampled	Yes	28–47
Rrs	Time	Peaks, minimum, Cr, Cl, E	No	54, 56, 61
Rrs	Time	20-point resampled	Yes	65–84
Flow	Time	Minimum	No	62
Flow	Time 2D	Landmark Cr,Cl, E	Yes	55, 57, 63–64
Flow	Time	20-point resampled	Yes	85–104
Rrs, Flow	2D	Landmark Z, B, A	Yes	48–53
Rrs, Flow	2D	Landmark Z and D	Yes	58–60
Rrs, Flow	2D	Mean and std of polar coordinators from 20-point Rrs, Flow	Yes	105–108
Rrs, Flow	2D Frequency	Maximum of full DWT from 20-point Rrs, Flow	Yes	109, 110
Rrs, Flow	Frequency	Maximum spectral coherence Rrs and Flow	Yes	111

5.5.1   Feature Pool

Landmark features are scalar values calculated from points of a breath cycle. Intuitively, we want to capture the boundary information of normal cycles to detect anomalies. For example, in Fig. 5.1a, A, B, CL, CR, D, E, F, Z are seven *landmark* points whose distances contain information for artefact detection (called *7-point extraction*). Specifically, points B and Z are at the zero flow value and the higher and lower Rrs values, respectively. Points A and D are at the maximum and minimum of Flow, respectively. Points CR, CL and E are at the maximum (right: positive Flow area and left: negative Flow area) and minimum of Rrs, respectively.

Resampling features are extracted from one dimensional input with a fixed number of points for a cycle to alleviate varied durations of breaths. We noticed that the minimum length of all breaths in the training data sets is larger than 30 points. For generalization, we consider 20 points per cycle and assume this is sufficient to describe the fundamental shape information for a breath curve. Thus, we re-sample Rrs, *Flow, Xrs, Volume* at a fixed rate of 20 points/cycle (called *20-point*) (Fig. 5.2).

Other new candidates in the pool are from different domains. For example, we obtain the changes of polar coordinates over time for each breath (using the mapping from Cartesian coordinates to their polar ones). We also explore the wavelet decom-

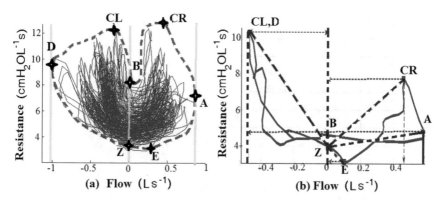

Fig. 5.1 **a**: All accepted breaths (*Rrs* against Flow) by specialists from one child (several recordings) and *7 points* proposed to determine *boundary landmarks* (dotted curves). **b**: Example features extracted by *landmarks* for one breath from a child (dotted lines: Euclidean distances between points)

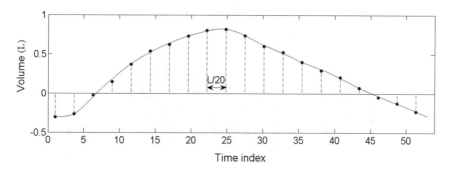

Fig. 5.2 Example of unified *20-point* resampling for a breath (Volume, time)

position analysis, DWT, (three level decomposition) with the Daubechies method [9] of the above 20-point resampling vectors. For the spectral coherence computation, we use 0.1667-second windows (as our frequency of interest is 6 Hz) and ensemble-average every three windows with 50% overlap. This and the impedance are then resampled at 10 Hz to effectively get the same number of coherence points as the number of impedance values. The existing are minima and maxima of *Rrs*, and DWT of *pressure* (e.g., [5, 6, 24, 26, 28]).

5.5.2 Challenging Factors and Other Criteria

Figures 5.3 and 5.4 illustrate the time dependence of samples within and between recordings (and between different age groups). These variations and artefacts are contained partly in the scaling information of the samples. This may introduce bias

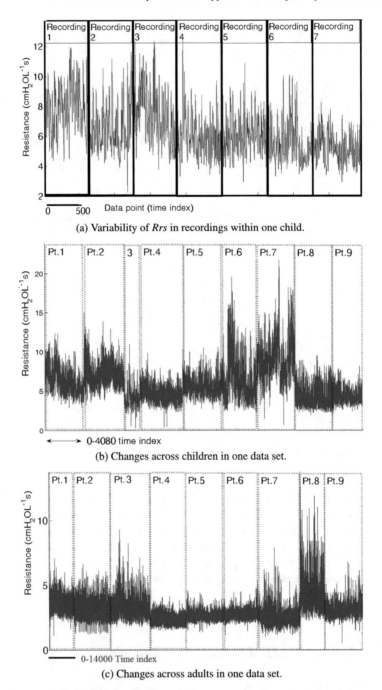

(a) Variability of *Rrs* in recordings within one child.

(b) Changes across children in one data set.

(c) Changes across adults in one data set.

Fig. 5.3 Examples of challenges in learning contaminated *Rrs* (after preprocessing) within a participant **a** and between participants **b, c**

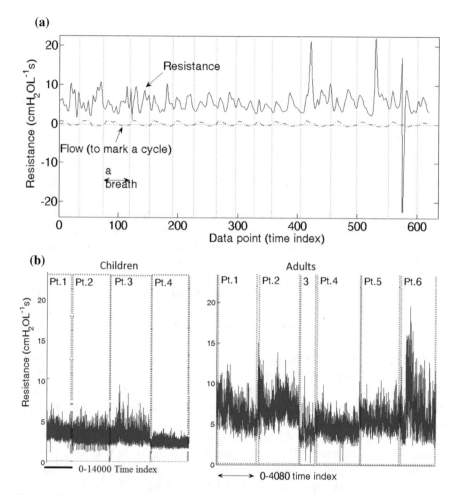

Fig. 5.4 Examples of challenges in learning contaminated *Rrs* (after preprocessing) within and between participants. **a**: Different breaths in one raw recording from a child. **b** Changes across children in one data set (left) and adults (right)

into ranking scores of features which are extracted from amplitude values across recordings. To reduce the bias, we accumulate scores for each feature candidate in a recording-wise manner.

Apart from saliency ranking, we select a relevant and efficient feature set based on performance metrics. We investigated ROC, F1-score, *throughput*, and *approval rate*. For clinical interest, we have quantified the reduction in artefactual activity and selected features by the variability of the average *Rrs*.

5.6   Unsupervised Artefact Detector

5.6.1   Single Filter Approach

In thresholding filters, a breath is marked as an artefact and discarded if one of its features exceeds a given upper bound or is less than a lower one. Since the normality hypothesis of *Rrs* in a recording is rejected with a significance level of 0.05 (the p-values were very close to zero; 0 to 1.27×10^{-17}) by the Lilliefors test [16] and the KS test [18], we do not assume a specific data distribution. Instead we use the ROC plots to determine the threshold parameters. We refer to this detector as a quartile thresholding filter.

Let Q_1, Q_3, and IQR denote the 25th, 75th percentiles and the interquartile range of a variable, respectively. Let n_{IQR} be a number of IQR intervals away from the Q_1 and Q_3. The lower bound θ_L is defined by n_{IQR} interquartile intervals less than Q_1. The upper bound θ_H is n_{IQR} intervals greater than Q_3, i.e.:

$$\theta_L = Q_1 - n_{IQR} \times \text{IQR} \tag{5.1}$$

$$\theta_H = Q_3 + n_{IQR} \times \text{IQR} \tag{5.2}$$

To simplify parameter settings, we apply the same n_{IQR} to all features and categorize subjects into two age groups (i.e., Paediatrics and Adults). We split each age group into two data sets: one for training and the other for test. In this work, the set of (training, test) for children is ($Ds1$, $Ds3$) and for adults is ($Ds2$, $Ds4$) (details of data sets are in Table 5.1).

We compare our detector with *B-3SD* [26] and the wavelet based with a complete-breath rejection approach [5] (namely *Wavelet-breath*). These two methods were recently proposed as the best automated ones in the literature. Note that, the work [5] used a point rejection approach and asked the participant to intentionally introduce artefacts while *Wavelet-breath* uses complete-breath rejection, and was tested with our real-life artefacts. We performed *Wavelet-breath* with three levels of DWT coefficients (cd1,cd2,cd3) and the *db5* method for *pressure*, and then used the three recommended thresholds [5] (i.e., $cd1^2 = 0.004$; $cd2^2 = 0.023$; $cd3^2 = 0.07$).

5.6.2   Multi-filter Approach

Utilising previous experimental observations [5, 22], we examine an additional case of employing the maximum of first level wavelet coefficients, $cd1$, decomposed from pressure. According to observations in reference [5] which used a data set with a predetermined plan to introduce artefacts, second and third level coefficients ($cd2$, $cd3$) were proposed to detect swallowing and leaks at the mouthpiece. These

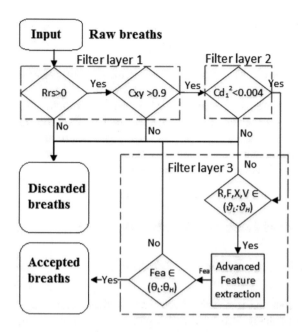

Fig. 5.5 Combined respiratory artefact detection scheme. *Rrs* is resistance values of input breaths. C_{XY} is the spectral coherence between pressure and flow values of breaths. $Cd1^2$ is the squared first level wavelet decomposition of pressure values. R, F, X, V and $(\Theta_L : \Theta_H)$ are resistance, flow, reactance, volume values and their corresponding threshold ranges. *Fea* and $(\theta_L : \theta_H)$ are features extracted (from the relationship between *Rrs* and *Flow* values) and their threshold ranges

artefacts often cause abnormal data points (i.e., out of usual range) in the *Rrs*-flow curves [26], therefore the associated features are expected to be able to detect them.

For *invisible* artefacts (e.g., light coughs) which could be in the usual range of normal breaths, the wavelet features of *cd*1 was found to be an alternative detection technique [5]. Hence, we investigate the contribution of this feature type as a separate layer of system for comparison purpose.

We used a three layer system for artefact removal. This comprises the pre-processing step, the wavelet decomposition step, and the interquartile range filter using landmark features *IQR-Landmark* (Fig. 5.5). We call this *IQR with landmark and wavelet (IQR-LW)*. Breaths that fail any threshold checking step are marked as artefacts and discarded (complete-breath approach). The remaining breaths after three filter layers are considered to be *clean* data (without artefacts).

The first layer (*Pre-processing*) is a *non-physiologically plausible denoise filter* that removes breaths containing data points which are physiologically implausible or corrupted by nonlinear noise using the FOT quality guidelines [17]. These include breaths containing negative *Rrs* values [26] or having magnitude-squared coherence values [7], equations as in Appendix A, C_{XY}, of pressure and flow less than 0.9 [17]. C_{XY} was calculated over 1/6-second windows, and ensemble-averaged every three windows with 50% overlap. This and the impedance were then resampled every 10 Hz to obtain the same number of coherence points as the number of impedance values.

After preprocessing, as its benefits shown in the previous works (discussed detail in the application background section) the squared first level of wavelet coefficient derived from pressure data is compared with a preset threshold of 0.004 $(cmH_2O)^2$. The second layer is expected to detect artefacts not identifiable by landmark features. In preliminary investigations (results not shown), we determined that the optimum performance was obtained using only $cd1$ (i.e., other coefficients can be ignored). The final step involves the *IQR-Landmark* filter.

5.7 Supervised Learning Artefact Detector

5.7.1 Machine Learning and Challenges

Figure 5.6 illustrates basic steps involved in a classification task, especially for supervised learning methods. The preprocessing module converts raw recorded pressure and flow into volume, *Rrs* and *Xrs* and removes non-eligible data. The feature extraction module transforms these variables into features mostly using the knowledge-based domain. Following modules are the newly proposed techniques to discriminate artefacts from normal breath cycles.

Several challenging factors for the classification task are the time-dependency of lung function (e.g., lung elasticity [20] as depicted in Fig. 5.3 and clinical aspects of FOT (e.g., *Rrs* and *Xrs* are dependent on body size and possibly racial/ethnic differences [30]). In Fig. 5.4, one can easily notice the difference of recordings between children and adults and even between different children.

These individual variations and physiological dependency may introduce bias into modelling, particularly when training samples need to be recruited from several different recording sessions. Because each FOT recording often lasts for one minute, thus there is only about twenty breaths within a recording. Moreover, within

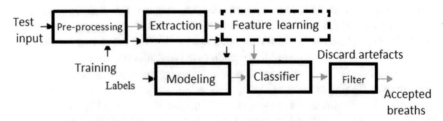

Fig. 5.6 Feature engineering scheme is a front-middle step of the applied machine learning process in which *feature learning* is used to *decorrelate* samples before using conventional general-purpose binary classifiers. Dark colour arrows for training stage; Light colour arrows for testing stage

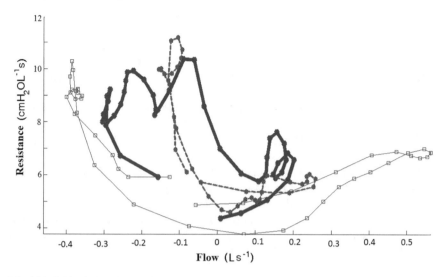

Fig. 5.7 Subjectivity of manual removal when regarding to as the ground truth (positives are artefacts). The line with round markers (increased line width) are real artefacts in one recording for a child that the operator mislabelled them as *accepted* during creating labels. The other two lines are two examples of correct manual labels in the same recording

a recording, samples also suffer from *correlation* which results from physiological time-dependence of the lung function [20]. These correlated training samples make modelling difficult. In this work, we introduce an intermediate module, called *feature learning*, that helps to *decorrelate* samples before using conventional general-purpose binary classifiers.

Another critical issue for supervised learning lies in subjectivity of labels. Human operators created labels based on recording-wise removal. Usually these labels that are used for training in machine learning are assumed as *ground truth*. However, we observed several cases that where the operator missed artefacts and marked them as *accepted* due to recording-wise removals and/or subjectivity of human. For example, Fig. 5.7 demonstrates wrong cases of False Positive evaluation results with regard to manual removal as the ground truth (positivesves are artefacts). This example is for one recording from a child when comparing results between the automated detector by our previous work [22] and the operator. In the figure, the traces with an increased line width were real artefacts in one recording for a child that the operator realised later when evaluating test methods (e.g. our previous work [22]) that he/she mislabelled them as *accepted* during labelling. Hence, conventional accuracies such as sensitivities and specificities may not reflect the true performance of the proposed method [22]. Thus, in this work, we report the evaluation using *approval rate* and *throughput* that are less dependent on manual labels than conventional accuracies but still depict the performance of the filtering function.

5.7.2 Feature Extraction

Though there is a big exploratory pool of features constructed from the knowledge domain (Sect. 5.5), we only investigate the selective sets made by saliency scores [22]. Specifically, we extract three top ten salient feature sets according to three ranking criteria: mutual information (MI) of feature candidates with artefacts measured by Shannon's information theory [29]; discrimination of features scored by Euclidean distances (DIS) [15]; and the variance ratio of clusters [1] (Var-Ratio) to measure the clusterability of features. Details of computing these scores can be found in Sect. 3.1.

As demonstrated in saliency ranking results [22], there are feature candidates that consistently being selected regardless of participant age and across all above three saliency criteria includes two dimensional extraction of *Rrs* and *Flow*, so-called *landmark*. Briefly, *landmarks* are *boundary* points (e.g. the zero flow point, the maxima and minima of flow and *Rrs*) of a breath cycle in a two dimensional presentation of *Rrs* and *Flow* [22].

5.7.3 Feature Learning and Supervised Classifier

The supervised learning classification proceeds in three steps:

1. Step 1: Gathering a training set, D_{tr}, and a test set, D_{te}.
2. Step 2: Feature extraction and *feature learning*.
3. Step 3: Modelling and classification.

In Step 1, from four separate datasets (i.e., Ds1, Ds2, Ds3, and Ds4) we build a study set $D = \{Ds1 \cup Ds2 \cup Ds3 \cup Ds4\} \stackrel{\text{def}}{=} \{D_{tr}, D_{te}\}$ where 70% of samples in D, called D_{tr} is used for training and 30% of samples in D, called D_{te} is used for out-of-sample tests.

In Step 2, after extracting the numerical features (Sect. 5.5.1), we extract *categorical* ones. This process aims to alleviate the aforementioned challenging factors (discussed in Sect. 5.5.2). We use Wilcoxon signed rank tests [36] to estimate the closest distribution i.e., the closet value of mean between an unseen test breath and a training group. Hence, the classifier that is used in the next step will operate in a batch-wise manner [35].

Specifically, consider a training set of N samples, $D_{tr} = \{s_i\}_{i=1:N}$. A sample $s \in D_{tr}$ has a numeric feature vector $\bar{x} \in R^d$, and its class label $y \in \{0, 1\}$. We transform x into an extended version that contains an additional vector $\bar{a}$ (namely a *de-correlating* feature vector) to cluster D_{tr} into nearly i.i.d groups. The motivation of using this component was discussed in detail at Sect. 5.5.2. Thus, learning inputs of a classifier will be $Feature = \{\bar{x}, \bar{a}, y\}_{i=1}^{N}$ where $\bar{a}$ is a function of $\bar{x}$.

Let $\bar{a}_{train}$ and $\bar{a}_{test}$ denote the $\bar{a}$ value of a sample in the training and test set, respectively. Given the total number of training participants, P, the maximum number of measurements per subject, M, and the maximum number of breaths per measurement, B, we identify s by using (p, m, b) where $p \in \{1, \ldots, P\}$, $m \in \{1, \ldots, M\}$, and $b \in \{1, \ldots, B\}$. $\bar{a}_{train}$ is defined by a function of p, m, b that consists of Kronecker delta functions $\delta(n - \alpha)$ where $L = P + M + B$ and $n \in \{1, \ldots, L\}$.

$$a_{(p,m,b)} = \delta(n - p) + \delta(n + P - m) + \delta(n + P + M - b) \qquad (5.3)$$

Since the number of training samples in each FOT measurement is very limited (about twenty samples per recording), the assumption of sufficient training samples [12] could not be applied. We assign $\bar{a}_{test}$ to the "most matched" $\bar{a}_{train}$, i.e., a group that the distribution of test sample is likely to belong to.

For time dependence, we gather breaths with the same ordinal number of a measurement into one group. If a test sample has a larger ordinal number than any training sample, it is given the largest ordinal number of the training. For difference of distributions, especially when P is small (i.e., only 20 in our case), the assignments use Wilcoxon signed rank tests [36]. The examination finds a training group in which its median is closest to the distribution of an unseen test sample. The null hypothesis in Wilcoxon tests is rejected with a significant of 0.05. The most matched is corresponding to the test case with the maximum p-value. To reduce an estimation error, each group is tested repeatedly within a range of one interquartile range from the training group (iteration steps of 1/200). We consider the p-value of Wilcoxon test that larger than 0.9 indicates a match.

In Step 3, we explore three different classifiers to compare the improvements made by our *feature learning*: *1-SVM*, *KNN*, and *Ensemble*. These classifiers are general-purposed classification algorithms with well-known implementations by MATLAB (The MathWorks Inc., Natick, MA, 2000) and LibSVM [8]. Detailed descriptions of these algorithms can be found [2, 14, 27]. In *1-SVM* tests, we use the Schkopf method [27] with the accepted breaths being the target class. We do a nested cross validation for model selection in *1-SVM*. The inner loop (10-fold) is considered a part of the model fitting procedure. The outer (5-fold) estimates the performance of this model fitting approach. A grid search for parameter ranges from 0.01 to 1 (steps of 0.05) using a radial basis kernel function. In *KNN* tests, we train a 5-nearest neighbors classifier. In *Ensemble* tests, we construct a boosted classification using the AdaBoost M1 method (with decision trees as the weak learners, 100 trees).

5.8 Results

5.8.1 Saliency Ranking

Ranking scores (i.e., DIS, MI, and Variance-Ratio) for each feature candidate are presented in Fig. 5.8. Features with circle markers have been currently used in the literature (Table 5.3) while the others are our new candidates. Group "I" illustrates results for children while "II" depicts adults. We sorted scores in the entire pool from high to low by each saliency criterion. Ranking order for the pool in Fig. 5.8a is from 1 to 111 (high to low); the higher saliency score indicates the higher ranking order. Figure 5.8b shows the top ten features with their identifications (IDs, detailed in Table 5.3).

As can be seen in Fig. 5.8a, only three previous features (the minimum and peaks of *Rrs*) are in the top ten highest ranking candidates from the children group. For adult cases, these *Rrs* features have moderate variance ratio and very low DIS scores. Our novel *landmark* features dominate not only in both children and adult groups but also across all three saliency criteria. Specifically, in Fig. 5.8b, they are *landmark* features ID 48, 49, 50, 54 56, and 64 (description for these features is in Table 5.3).

In next steps, given a detector of interest, we continue the selection by performance criteria.

5.8.2 Unsupervised Artefact Detector: Single Filter Results

5.8.2.1 Parameter Settings

Using a quartile thresholding detector, against a wide range of deviation threshold parameters, we compared the ROC, F1-scores, *throughput*, and *approval rate* curves (Fig. 5.9) and the variability (Fig. 5.10) of three selection schemes with the case of no selection. Apart from examining effects of introducing the selection schemes, we use the above curves to determine the optimized n_{IQR} settings.

We explored n_{IQR} in a range $0 \rightarrow 10$ (incremental steps of 0.25). For $n_{IQR} > 4$, curves did not vary significantly. Hence, we depict these curves only for $n_{IQR} \leq 4$. Four criteria (*DIS, MI, Variance Ratio* or *No-sel* (i.e., no selection)) are presented with different markers. Figure 5.9a and e presents their F1-scores. ROCs are shown in Fig. 5.9b and f with solid lines for sensitivity, and dotted for specificity. The *throughput* and clinical *approval* rate of the removal are illustrated in Fig. 5.9c and g, d and h, respectively. The effects of n_{IQR} and the feature selection on the variability are demonstrated in Fig. 5.10.

We observed that characteristic curves were different between age groups. When no feature selection algorithm is used, the optimized empirical n_{IQR} is 3 for children and 2 for adults. If a feature selection algorithm is involved, the optimized empirical n_{IQR} reduced to around 1.5 for children or 1 for adults. F1-score and *throughput* are also improved significantly.

One important parameter setting is $n_{IQR} = 1$. Across three feature selection algorithms, this setting can work in a subject-independent manner with a high sensitivity (around 80%) and specificity (about 70%) regardless of participant age (i.e., children or adults). Although curves of the three saliency criteria were quite comparable, in *approval rate* and variability, the MI selection is better. Hence, we proposed a final model for the quartile thresholding detector that uses the MI selection technique and settings of $n_{IQR} = 1$, called *1IQR-MI*. In the next section, we do out-of-sample

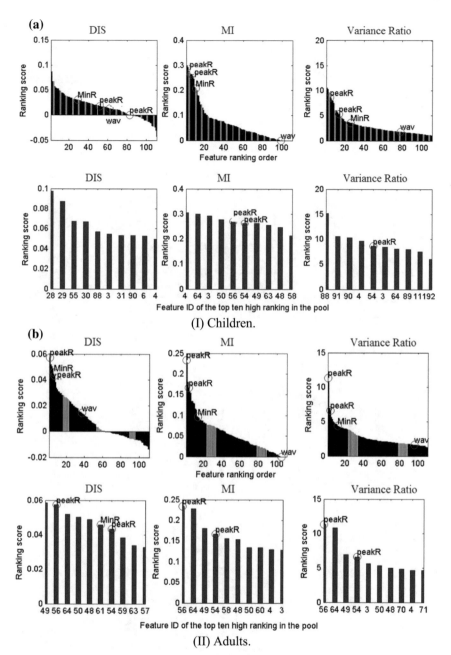

Fig. 5.8 Ranking scores for the feature pool (**a**) and the ten highest-score candidates (**b**). Vertical axes: scores calculated by three saliency criteria. Horizontal axes in **a**: ranking order (highest = 1, lowest = 111); **b**: feature identification (ID) in the pool. Circle markers: existing features (details in Sect. 5.5.1). Note that *peakR* annotations in plots indicate a peak at either the left or right handside of a Rrs-flow curve

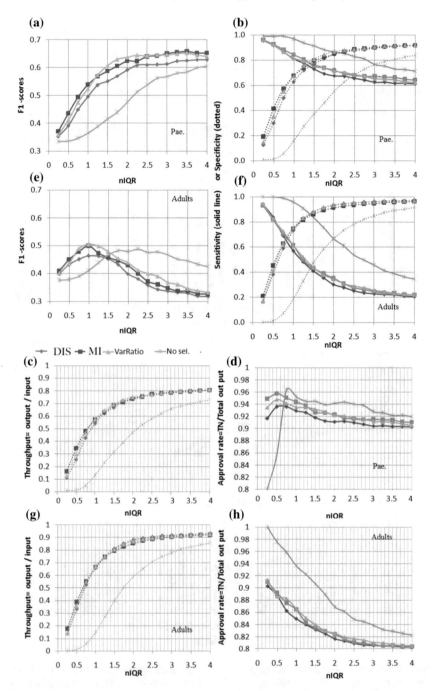

Fig. 5.9 Effects of n_{IQR} and feature selection for paediatrics (top) and adult (bottom). Markers are for different feature selection algorithms. **a, e** are for F1-scores. **b, f** are for ROC curves (Solid lines: sensitivity; the dotted: specificity). **c, g** are for *throughput* curves. **d, h** are for *approval rate*

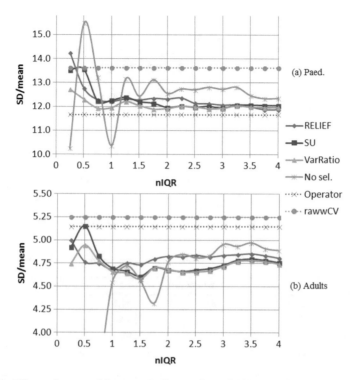

Fig. 5.10 Effects of n_{IQR} and feature selection on the variability of the average Rrs (standard deviation over the mean across patients). Markers are for different selection algorithms. **a** is for children and **b** is for adults

test with this model and compare with the aforementioned existing artefact removal methods.

5.8.2.2 Out-of-Sample Tests

We used unseen test sets ($Ds3$ for children and $Ds4$ for adults) to validate the proposed detector ($1IQR$-MI). Table 5.4 compared $1IQR$-MI with existing complete-breath based methods: B-$3SD$ [26] and *Wavelet-breath* [5]. *Manual* is the reference value calculated from removals by a human expert. Paired t-tests (two-tailed) for the variability (the test minus the operator, degrees of freedom of four ($Ds3$) or nine ($Ds4$)) are also reported in Table 5.4. In terms of sensitivity, approval rate by operator (i.e., of the output are breaths "accepted" by the clinician), and the variability, $1IQR$-MI is the best detector. For example, in adults, although the mean Rrs of the $1IQR$-MI had a comparable average value with the operator, the standard deviation is lower (only 0.40 while the operator was 0.44 with p value is 0.06).

Table 5.4 Results of out-of-sample test phase. *IIQR-MI*[a] is the proposed. Others are the existing. *P* values[b] are from paired t-tests (two-tailed, $N = 5$ (*Ds3*) or 10 (*Ds4*)). Positives are artefacts[c]. Performance metrics are percentages. The proposed *IIQR-MI* has the highest approval rate and the closet variation of standard deviation of *Rrs* to the manual labelling

	Paediatrics (test *Ds3*)			
	B-3SD [26]	Wavelet-breath [5]	IIQR-MI[a]	Manual
F1-score[c] [%]	46.8	21.6	41.2	–
Approval[c] [%]	95.4	94.8	98.0	–
Throughput[c] [%]	82.7	30.0	67.1	84.1
Mean($\pm$SD) *Rrs* [cmH_2OsL^{-1}]	3.72 ($\pm$0.18)	3.74 ($\pm$0.20)	3.70 ($\pm$0.17)	3.75 ($\pm$0.16)
P-value[b] *Rrs*	0.08	0.84	0.03	–
Mean($\pm$SD) SD*Rrs* [cmH_2OsL^{-1}]	0.29 ($\pm$0.13)	0.38 ($\pm$0.12)	0.32 ($\pm$0.11)	0.32 ($\pm$0.13)
P-value[b] SD*Rrs*	0.23	0.25	0.82	–
	Adults (test *Ds4*)			
F1-score[c] [%]	50.6	49.3	54.7	–
Approval[c] [%]	77.4	78.1	80.6	–
Throughput[c] [%]	85.4	55.9	63.4	68
Mean($\pm$SD) *Rrs* [cmH_2OsL^{-1}]	3.69 ($\pm$0.97)	3.69 ($\pm$0.98)	3.66 ($\pm$0.94)	3.67 ($\pm$0.95)
P-value[b] *Rrs*	0.34	0.58	0.14	–
Mean($\pm$SD) SD*Rrs* [cmH_2OsL^{-1}]	0.40 ($\pm$0.23)	0.41 ($\pm$0.27)	0.40 ($\pm$0.24)	0.44 ($\pm$0.21)
P-value[b] SD*Rrs*	0.05	0.86	0.03	–

[a]The detector used one interquartile range as a subject-independent parameter with the top ten salient features selected by the MI technique
[b]compared to *Manual*, significant if $P < 0.05$
[c]Removals by a specialist is considered ground truth. *Throughput* is the ratio of breath numbers in the output to input. *Approval rate* of the breaths remained after removal is the ratio of breaths that are "accepted" by the human to total output breaths. Details of equations as in Sect. 5.3

Rrs in our study ranged from $1.7 \rightarrow 8\ cmH_2OL^{-1}s$ in adults (Table 5.4), i.e. a mild to medium range of obstruction. To investigate the potential influence of obstruction on our detector, Fig. 5.11a shows one performance metric, i.e. the approval rate, plotted against the median resistance for each recording, while Fig. 5.11b shows the distribution of the approval rate. It can be seen that while there is a large range, the approval rate remains mostly high regardless of the median *Rrs*. Similarly in children, Fig. 5.11c and d show that, with the exception of three recordings, approval rates remain high regardless of median *Rrs*, albeit within a smaller range of resistances. We also quantified the independence of the approval rate and *Rrs* using the distance

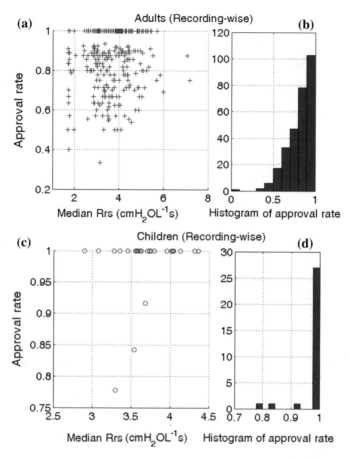

Fig. 5.11 Approval rate plotted against median *Rrs* and histogram of approval rates for all recordings for the adults (**a, b**) and children (**c, d**) testing datasets

correlation [31], and obtained 0.12 for adults and 0.27 for children, with complete independence indicated by 0.

5.8.3 Unsupervised Artefact Detector: Multi-filter Results

5.8.3.1 Comparisons of Filters Against Ground Truth

In terms of comparison against the manual operator as ground truth, Fig. 5.13 presents the receiver-operator characteristic of the proposed filter across a range of n_{IQR} values for both adult and paediatric data. We found that $n_{IQR} = 1.5$ gave the best performance in adult data, whereas $n_{IQR} = 2.5$ gave the best performance for paediatric data. An

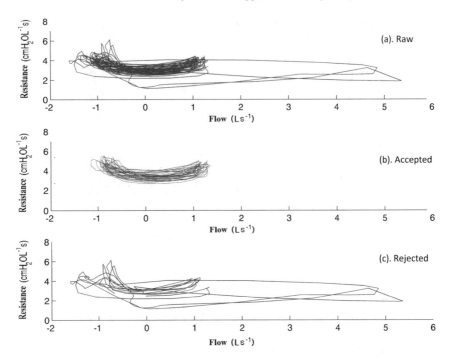

Fig. 5.12 Example of filtering a measurement from an adult by our proposed method. **a**: the unfiltered data (i.e. contains artefacts). **b**: the accepted data or output of the system. **c**: the discarded data by the system

example of filtering for a measurement is illustrated in Fig. 5.12 for an adult with $n_{IQR} = 1.5$ (Fig. 5.13).

With the best performing n_{IQR} values, the combined method achieved 76% (adult) and 79% (paediatric) agreement with the manual operator. The performance metrics for the filters studied are shown in Table 5.5. Note that since the manual operator labelled acceptability in terms of breaths and not points, metrics are not available for the wavelet-point method.

5.8.3.2   Comparisons Between Filters

Tables 5.6 and 5.8 show the variability of filtered Rrs profiles across test methods in comparison with the unfiltered data and filtering by a manual operator, for the training and test datasets, respectively.

The percentage of breaths that were removed by the first layer from raw data sets were only about 1% (paediatric) and 2% (adult) (Table 5.6). The remaining breaths that were kept by our method is 69% (paediatric) and 73% (adult) of the total raw input (the manual method kept about 77% in both cases). While the *Wavelet-point* method

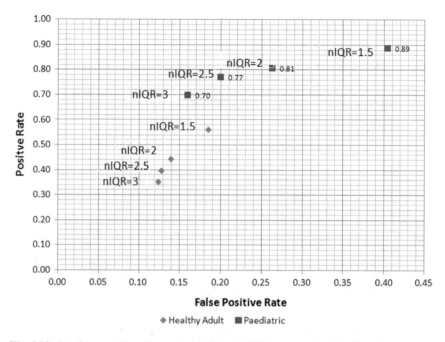

Fig. 5.13 Receiver operating characteristic of the multi-filter approach with adjustable parameter n_{IQR} (range from 1.5 to 3) for paediatric group (square) and adult group (diamond)

Table 5.5 Comparisons of filters against the manual operator during training. *IQR-Landmark*[a] and *IIQR-MI*[b] are our works related to our current proposed, *IQR-Combined*[c]. textitWavelet-breath[d] is the existing. Positives are artefacts. True positive breaths are breaths rejected by both machine-based and manual removal. F1-score[e] is the harmonic mean of precision and sensitivity. Sens: Sensitivity. Spec: Specificity

Method	Healthy adults				Paediatrics			
	Accuracy[e]	F1[e]	Sens[e]	Spec[e]	Accuracy[e]	F1[e]	Sens[e]	Spec[e]
IQR-Landmark[a]	0.753	0.545	0.640	0.787	0.693	0.525	0.842	0.655
Wavelet-breath[d]	0.584	0.335	0.453	0.623	0.431	0.341	0.730	0.356
IIQR-MI[b] [22]	0.763	0.571	0.683	0.787	0.731	0.553	0.824	0.708
IQR-Combined[c]	0.781	0.569	0.626	0.828	0.827	0.632	0.734	0.851

[a] A single filter approach with landmark features and $nIQR = 1.5$ for adults and 2.5 for children (where relevant)
[b] A single filter approach with features automatically selected by ranking [22] and $n_{IQR} = 1$ for both age groups
[c] A multi-filter approach (comprising a wavelet and *IQR-Landmark*) with $n_{IQR} = 1.5$ (adults) or 2.5 (children)
[d] A complete breath rejection approach using the wavelet coefficient thresholding detecion [5]
[e] Removals by a specialist is considered ground truth

Table 5.6 Comparison of filtered *Rrs* profiles between filters during training. *IQR-Landmark*[a] and *IIQR-MI*[b] are our works related to our current proposed, *IQR-Combined*[c]. *Wavelet-point* and *Wavelet-breathare*[d] the existing. *wCV* and *bCV* are within and between session coefficients of variation (Sect. 5.3) and presented in %. *P* values[e] are from paired t-tests (two-tailed). $\%_{out}$ is the percentage of remaining *breaths* (against the total raw input, unit in %) after being filtered by methods except for *Wavelet-point* which is in percentage of the raw data points. $\%_{\text{discarded-by-preprocessing}}$ is the percentage of artefacts that were removed in the preprocessing step (a common step for all test filters)

Method	Healthy adults					Paediatrics		
	wCV	P-value wCV[e]	**bCV**	P-value bCV[e]	$\%_{out}$	**wCV**	P-value wCV[e]	$\%_{out}$
Unfiltered (raw data)	*5.25*	–	*6.69*	–	*100.0*	*13.62*	–	*100.0*
Manual (reference)	**5.14**	–	**6.31**	–	**76.9**	**11.66**	–	**77.2**
IQR-Landmark[a]	4.56	0.08	5.76	0.18	80.6	12.69	0.57	74.5
Wavelet-point [5]	5.43	0.34	6.84	0.46	97.1	13.96	0.30	98.9
Wavelet-breath[d]	5.93	0.20	7.82	0.34	98	11.9	0.85	77.8
IIQR-MI[b]	4.69	0.20	5.91	0.05	67.8	12.25	0.80	60.0
IQR-Combined[c] (proposed)	4.57	0.11	5.75	0.17	72.8	13.27	0.32	69.6
$\%_{\text{discarded-by-preprocessing}}$					1.9	2.6		

[a]A single filter approach with landmark features and $nIQR = 1.5$ for adults and 2.5 for children (where relevant). and *IIQR-MI*
[b]A single filter approach with features automatically selected by ranking [22] and $n_{IQR} = 1$ for both age groups
[c]A multi-filter approach (comprising a wavelet and *IQR-Landmark*) with $n_{IQR} = 1.5$ (adults) or 2.5 (children)
[d]A complete breath rejection approach using the wavelet coefficient thresholding detecion [5]
[e]Removals by a specialist is considered ground truth

kept 99% (paediatric) and 97% (adult) of total raw data points, *Wavelet-breath* only kept 78% (paediatric) and 98% (adult) of raw breaths. Without the wavelet layer, *IQR-Landmark* produced 74% (paediatric) and 81% (adult).

For a completely out-of-sample test set (adult patients with respiratory disease), the above performance was maintained (Tables 5.7 and 5.8). Our method kept 66% of breaths compared with 69% of the human method. The accuracy of children test set is 89.1%, higher than 82.7% of the training performance.

Table 5.7 Comparisons of filters against the manual operator with out-of-sample data. *IQR-Landmark[a]* and *1IQR-MI[b]* are our works related to our current proposed, *IQR-Combined[c]*. *Wavelet-breath[d]* is the existing. Positives are artefacts. True positive breaths are breaths rejected by both machine-based and manual removal. F1-score[e] is the harmonic mean of precision and sensitivity. Sens:Sensitivity. Spec: Specificity

Method	Asthma adults				Paediatrics			
	Accuracy[e]	F1[e]	Sens[e]	Spec[e]	Accuracy[e]	F1[e]	Sens[e]	Spec[e]
IQR-Landmark[a]	0.719	0.610	0.715	0.720	0.738	0.398	0.848	0.725
Wavelet-breath[d] [5]	0.596	0.435	0.506	0.636	0.369	0.179	0.674	0.334
1IQR-MI[b] [22]	0.736	0.606	0.661	0.769	0.747	0.412	0.870	0.733
IQR-Combined[c]	0.731	0.609	0.683	0.752	0.891	0.588	0.761	0.906

[a] A single filter approach with landmark features and $nIQR = 1.5$ for adults and 2.5 for children (where relevant)

[b] A single filter approach with features automatically selected by ranking [22] and $n_{IQR} = 1$ for both age groups

A multi-filter approach (comprising a wavelet and *IQR-Landmark*) with $n_{IQR} = 1.5$ (adults) or 2.5 (children)

[c] A complete breath rejection approach using the wavelet coefficient thresholding detecion [5]

[d] Removals by a specialist is considered ground truth

5.8.4 Supervised Artefact Detector: Machine Learning Classifier Results

We evaluated the performance gain of our new proposed modules, feature engineering with selection and learning, across classifiers (*1-SVM, KNN, Ensemble*) in the classification task. For each classifier, we also implemented three feature selection algorithms: DIS, MI, Var-Ratio and the case without using selection (i.e. using all feature candidates, called No-Sel). Figure 5.14 demonstrates examples for out-of-sample tests with adult data.

As can be seen in Fig. 5.14, using our feature engineering module (referred to as "with") results in a higher *Approval* rate (i.e., positive difference) than the case without the module (referred to as "without"). Note that in clinical applications, an improvement of a few percentage in accuracy is very precious. Intuitively, compared with the "without", our automated machine learning based method using the proposed feature engineering module reported more TNs and less FNs, i.e., closer to the groundtruth.

We also notice that the proposed feature engineering module unfavorably has lower F1-score than the "without" case under the DIS, Var-Ratio feature selection criteria as well as No-Sel. However, if combining with the *throughput* metric, this higher F1-score under "without" is likely linked to the larger throughput difference between "without" and the manual method, especially the No-Sel case (about 42%).

Table 5.8 Comparisons between filters during out-of-sample tests using the *Rrs* profile (reference: manual work (highlighted row),) *IQR-Landmark*[a] and *1IQR-MI*[b] are our works related to our current proposed, *IQR-Combined*[c]. *Wavelet-point* and *Wavelet-breath*[d] are the existing. *wCV* and *bCV* are in %. P values[e] are from paired t-tests (two-tailed). $\%_{out}$ is the percentage of remaining *breaths* (against the total raw input, unit in %) after being filtered by methods except for *Wavelet-point* which is in percentage of the raw data points. $\%_{discarded-by-preprocessing}$ is the percentage of artefacts that were removed in the preprocessing step (a common step for all test filters)

Method	Asthma adults					Paediatrics		
	wCV	P-value wCV[e]	**bCV**	P-value bCV[e]	$\%_{out}$	**wCV**	P-value wCV[e]	$\%_{out}$
Unfiltered (raw data)	6.25	–	7.95	–	100.0	8.41	–	100.0
Manual (reference)	**6.86**	–	**8.86**	–	**68.9**	**8.55**	–	**89.8**
IQR-Landmark[a]	6.52	0.13	8.22	0.05	80.6	8.30	0.66	74.5
Wavelet-point [5]	6.64	0.57	8.15	0.09	98.7	9.06	0.21	79.1
Wavelet-breath[d]	7.51	0.37	8.35	0.24	97.9	10.12	0.23	33.3
1IQR-MI[b]	7.93	0.26	6.68	0.05	63.7	8.62	0.86	67.1
IQR-Combined[c] (proposed)	6.46	0.12	8.18	**0.03**	65.6	8.22	0.62	66.9
$\%_{discarded-by-preprocessing}$					2.5	0.9		

[a] A single filter approach with landmark features and $nIQR = 1.5$ for adults and 2.5 for children (where relevant)
[b] A single filter approach with features selected by ranking [22] and $n_{IQR} = 1$ for both age groups
[c] A multi-filter approach (comprising a wavelet and *IQR-Landmark*) with $n_{IQR} = 1.5$ (adults) or 2.5 (children)
[d] A complete breath rejection approach using the wavelet coefficient thresholding detection [5]
[e] compared to *Manual operator* significant if $P < 0.05$

On the other hand, our proposed method not only yields a higher F1-score for the MI feature selection criterion but also has the smallest throughput difference with respect to the manual method.

Combining all three performance metrics, we find that among implemented feature selection criteria (MI, DIS, Var-Ratio and No-Sel), the MI criterion yields the best performance for all implemented classifiers (*1-SVM, KNN, Ensemble*). From the experimentation, classifiers equipped with the feature learning and MI selection had a higher approval rate with children data than with adults. Meanwhile, these classifiers had a closer throughput to the manual output for adult than children.

Fig. 5.14 Effects of feature selection algorithms (DIS, MI, and Var-Ratio) and feature learning across classifiers during out-of-sample tests for adults. Difference in % (with standard errors) of *Approval*, *throughput*, and F1-score are comparisons between with and without feature learning. MI group produced gains in F1-score and the lowest difference in *throughput*

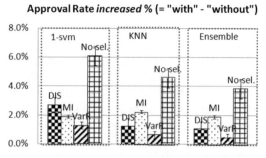

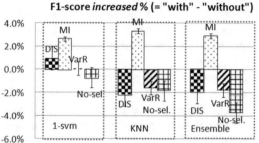

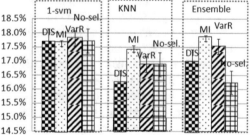

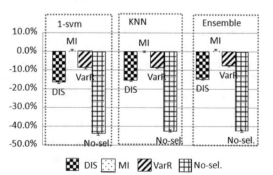

5.9 Discussion

Our experiments were executed on recordings collected from adults and eight- to eleven-year-old children in Queensland and New South Wales, Australia. For the feature extraction, we suggest to obtain *landmark* features of the two dimensional resistance-against-flow curves. This feature group is highly ranked by supervised learning techniques using saliency scores (DIS, MI, variance ratio). Given a training set, we calculated mutual information (Shannon's information theory) and cluster-ability scores to search for the best features. The MI score measures the correlation (mutual information) between one feature candidate of a breath and its label of abnormality. Meanwhie, DIS and variance ratio scores depict the clusterability of a feature candidate. The selective features we introduced result from analysis of saliency scores that has been motivated from a solid theoretical foundation and have a unified framework.

Although selecting the ten highest score candidates is common practice in the literature of feature learning, we acknowledge that an investigation for the stability of these feature preferences should be undertaken. Nevertheless, our results are consistent with more than one well-known feature selection algorithm with four separate data sets. As can be seen in Figs. 5.8, scores that lie out of the top ten were significantly lower than the highest level. Thus, $k = 10$ satisfied our requirements. In practice, one may choose the entire landmark group and the resulting detector will perform comparably to the approach of this work. This is because the majority of the top ten percentage are actually landmark features and the performance curves varied negligibly among selection algorithms.

We noticed that the normality hypothesis of *Rrs* in a recording was rejected with a significance level of 0.05 (the p-values were very close to zero; 0 to 1.27×10^{-17}). Hence, rather than assuming the normality of measurements and fixed threshold values (e.g. $3SD$ away from the mean) as in earlier works, we advised to use quartile percentages to detect anomalies, and measurement statistics to consider a breath an artefact if one of its features exceeds a given upper bound or is less than a lower one.

In the past, quality control of forced oscillation data has often been done on the basis of measures such as coherence, i.e. the degree of correlation between the oscillatory flow and pressure waves, where coherence values less than 0.95 were typically excluded [17]. However, this has known limitations: coherence is highly dependent on windowing and other signal processing settings for impedance calculation, and it is often much reduced in disease, especially at lower frequencies.

Compared with either 3SD or 5SD filtering (i.e., existing statistical filters based on number of standard deviations from the mean *Rrs* or *Xrs* value [6, 28]), we found that breath-based filtering resulted in lower within- and between-session variability in children. We also proposed removal of transient artefacts based on the distinct deviations observed in the oscillatory flow and admittance signals, and in the *Rrs*-flow profile [26]. Specifically, mouthpiece leak artefacts manifest as a marked increase in oscillatory flow and a pronounced spike in the magnitude of admittance. Other artefacts often contain depressions or gaps in the oscillatory flow signal but are best identified by examining the *Rrs*-Flow profile (e.g. spikes in *Rrs* at or near zero flow) [10, 24, 26]. However, these observations were made subjectively, with no quantitative criteria or threshold to determine exclusion. The results of the present study represent a first step towards more objective and automated criteria for quality control of FOT measurements, based on a complete breath strategy. It employs an intuitive approach to detecting anomalies from the *Rrs*-flow profile, for the first time using landmark features to identify outliers.

A more recent approach to artefact detection was using wavelet decomposition applied to the pressure profile of the breath [5], which was effective at excluding light coughing, swallowing and vocalization artefacts. Although the wavelet method has high performance in sensitivity and specificity (over 90%), its evaluation was limited to simulated artefacts by trained subjects, and its performance on real world data was unknown. In this study, using retrospective clinical FOT data, we found that partially incorporating the wavelet approach into our proposed algorithm, particularly that component which detects artefacts invisible to the operator from the FOT recording, resulted in superior performance compared to either method alone.

We found that the method performed best when partly combined with the previously published wavelet detection method. We tested the different filtering methods using real data collected from a variety of subjects: children, healthy and asthmatic adults. A high degree of agreement between our method and the manual work was observed and several breaths containing artefacts missed by the manual operator were detected by our method. Finally, within- and between-session variability was used to assess the performance of each filtering method in the absence of ground truth. The combined method reduced both variabilities compared with the operator, with a slightly higher exclusion rate. Though using the *IQR-Landmark* scheme produced a similar variation, a much lower exclusion rate than the operator implies that it may have missed several artefacts that were recognized by the human (Table 5.9).

The importance of feature engineering in applied machine learning is also presented via respiratory artefact removal in lung function tests. Specifically, we showed examples of challenges associated with individual variation and physiological dependency of samples when applying conventional general-purpose binary classifiers. We developed a *feature learning* module in order to overcome these challenging factors. The *feature learning* module decorrelates breaths and increases the detection performance of the classifiers. Our experiments were executed on four datasets collected from both adults and children in different site locations (15 children + 9 healthy adults + 10 adults with asthmatic; producing 470 FOT recordings that contains 8704 breath cycles). We use 70% of these datasets for training and 30% for out-of-sample tests. In total, we trained general-purpose classifiers with 6926 breath cycles (samples) consisting of 5518 normal breaths and 1408 artefacts. Among popular feature selection criteria in the literature (mutual information, Euclidean distance, and variance ratio of clusters), our feature engineering steps significantly improve performance of all implemented classifiers (*1-SVM*, *KNN*, *Ensemble*) with feature inputs selected by mutual information criterion.

Table 5.9 Test confusion matrix for out-of-sample tests with MI selection and feature learning. Positives are artefacts[a]. 1-SVM[b], KNN[c] and *Ensemble*[d] were compared with *Manual* regarding *throughput*

Classifiers	Paediatrics		Adults	
	Approval[c]	Throughput[a]	Approval[a]	Throughput[c]
1-SVM[b]	91.4	53.4	71.5	63.3
KNN[c]	91.4	53.4	71.6	62.8
Ensemble[d]	91.9	53.1	71.3	63.5
Manual[e]	–	85.3	–	62.8

[a]Removals by a specialist is considered ground truth. *Approval rate* [%] is the ratio of breaths that are "accepted" by the human to the total output breaths. *Throughput* [%] is the ratio of breath numbers in the output to input
[b]is one class SVM using accuracy as the cost function and a nested validation with 10-fold inner and 5-fold outer; A grid search for parameter ranges from 0.01 to 1 (steps of 0.05) using a radial basis kernel function
[c]5-nearest neighbors classifier
[d]A boosted classification using the AdaBoost M1 method (with decision trees as the weak learners, 100 trees)

5.10 Summary

This chapter presents a collective anomaly detection application using feature engineering proposed in Chap. 3. Experiments were executed on recordings collected from adults and eight- to eleven-year-old children in Queensland and New South Wales, Australia. Several novel feature extraction methods are suggested. Three approaches of detection schemes were investigated: unsupervised filters with single or multi-layers and supervised machine learning classifiers. We propose the work of multi-layer unsupervised filters for its best performance and low computation cost. In out-of-sample tests, this detector performed similar to the *gold standard*, as assessed by paired t-tests (two-tailed) for variability.

Lack of standardization in quality control of FOT has been a barrier to its widespread clinical adoption, despite decades of studies showing promising physiological and clinical relevance. The ability to remove common artefacts using objective and automatable criteria is critical to overcoming this barrier, as these approaches can be eventually incorporated into commercial software to guide the user and minimize inter-operator variability. These approaches are also especially desirable in emerging applications of FOT such as in epidemiological field testing [13] and home monitoring [10, 33].

There are few limitations in this work for the scenario. We only used data recorded for a single frequency of FOT (6 Hz) closest to what is commonly reported in the literature (5 Hz). However, our scheme could indeed be applied to multi-frequency systems, either treating each component frequency independently (where detection of an artefact at any frequency would result in the exclusion of a breath), or by feeding the most relevant features from all frequencies into the detector. The applicability of the detector at other frequencies remains to be tested, but we know that the Rrs-flow profiles appear similar at 6, 11 and 19 Hz such that the landmark features would likely be relevant.

When evaluating machine learning classifiers, only general-purpose well-known classifiers are examined. We suggest that taking advantange of our feature engineering result, a specialized supervised learning algorithm should be developed as future works to further improve the performance of artefact removal process.

In terms of applicability, the test datasets we examined exhibited a mild to medium range of obstruction, ranging in Rrs from 1.7 to 8 cmH_2OsL^{-1}. Thus our method will need to be tested for applicability across a wide range of obstruction, e.g., severely obstructed patients or during an exacerbation. However, we note that our performance metrics remained mostly high (approval rate $\geq 75\%$) regardless of median Rrs in both the children and adult datasets. There was also a low correlation between approval rate and Rrs as reported previously [22]. Further work will also be needed to determine how our method will perform in other diseases, e.g., chronic obstructive pulmonary disease, where abnormalities in reactance (Xrs) will likely dominate those in Rrs. For this, a new set of top ranking features may first need to be determined in a training dataset.

References

1. Ackerman M, Ben-David S (2009) Clusterability: a theoretical study. In: International conference on artificial intelligence and statistics, pp 1–8
2. Altman NS (1992) An introduction to kernel and nearest-neighbor nonparametric regression. Am Stat 46(3):175–185
3. Bateman E, Hurd S, Barnes P, Bousquet J, Drazen J, FitzGerald M, Gibson P, Ohta K, O'byrne P, Pedersen S et al (2008) Global strategy for asthma management and prevention: gina executive summary. Eur Respir J 31(1):143–178
4. Beydon N, Davis S, Lombardi E, Allen J, Arets H, Aurora P, Bisgaard H, Davis G, Ducharme F, Eigen H, Gappa M, Gaultier C, Gustafsson P, Hall G, Hantos Z, Healy M, Jones M, Klug B, Carlsen K, McKenzie S, Marchal F, Mayer O, Merkus P, Morris M, Oostveen E, Pillow J, Seddon P, Silverman M, Sly P, Stocks J, Tepper R, Vilozni D, Wilson N (2007) An official American thoracic society/European respiratory society statement: pulmonary function testing in preschool children. Am J Respir Crit Care Med 175(12):1304–1345. https://doi.org/10.1164/rccm.200605-642ST
5. Bhatawadekar SA, Leary D, Chen Y, Ohishi J, Hernandez P, Brown T, McParland C, Maksym GN (2013) A study of artifacts and their removal during forced oscillation of the respiratory system. Ann Biomed Eng 41(5):990–1002
6. Brown NJ, Thorpe CW, Thompson B, Berend N, Downie S, Verbanck S, Salome CM, King GG (2004) A comparison of two methods for measuring airway distensibility: nitrogen washout and the forced oscillation technique. Physiol Meas 25(4):1067–1075
7. Challis R, Kitney R (1990) Biomedical signal processing (part 3 of 4): the power spectrum and coherence function. Med Biol Eng Comput 28(6):509–524
8. Chang CC, Lin CJ (2011) Libsvm: a library for support vector machines. ACM Trans Intell Syst Technol 2(3):27:1–27:27
9. Daubechies I, Bates BJ (1993) Ten lectures on wavelets. J Acoust Soc Am 93(3):1671–1671

10. Dellacà RL, Pompilio P, Walker P, Duffy N, Pedotti A, Calverley PM (2009) Effect of bron-chodilation on expiratory flow limitation and resting lung mechanics in COPD. Eur Respir J 33(6):1329–1337
11. DuBois AB, Brody AW, Lewis DH, Burgess BF et al (1956) Oscillation mechanics of lungs and chest in man. J Appl Physiol 8(6):587–594
12. Dundar M, Krishnapuram B, Bi J, Rao RB (2007) Learning classifiers when the training data is not IID. In: Proceedings of the 20th international joint conference on artifical intelligence, IJCAI 2007. Morgan Kaufmann Publishers Inc., San Francisco, USA, pp 756–761. http://dl.acm.org/citation.cfm?id=1625275.1625397
13. Ezz WN, Mazaheri M, Robinson P, Johnson GR, Clifford S, He C, Morawska L, Marks GB (2015) Ultrafine particles from traffic emissions and children's health (UPTECH) in Brisbane, Queensland (Australia): study design and implementation. Int J Environ Res Public Health 12(2):1687–1702, https://doi.org/10.3390/ijerph120201687. https://eprints.qut.edu.au/82503/
14. Freund Y, Schapire RE (1997) A decision-theoretic generalization of on-line learning and an application to boosting. J Comput Syst Sci 55(1):119–139
15. Kira K, Rendell LA (1992) The feature selection problem: traditional methods and a new algorithm. In: Proceedings of the 10th national conference on artificial intelligence, AAAI 1992. AAAI Press, pp 129–134. http://dl.acm.org/citation.cfm?id=1867135.1867155
16. Lilliefors HW (1967) On the Kolmogorov-Smirnov test for normality with mean and variance unknown. J Am Stat Assoc 62(318):399–402
17. Lorino H, Mariette C, Karouia M, Lorino A (1993) Influence of signal processing on estimation of respiratory impedance. J Appl Physiol 74(1):215–223
18. Massey FJ Jr (1951) The Kolmogorov-Smirnov test for goodness of fit. J Am Stat Assoc 46(253):68–78
19. Mazaheri M, Clifford S, Jayaratne R, Megat Mokhtar MA, Fuoco F, Buonanno G, Morawska L (2013) School childrens personal exposure to ultrafine particles in the urban environment. Environ Sci Technol 48(1):113–120
20. Nunn JF (2013) Applied respiratory physiology. Butterworth-Heinemann
21. Oostveen E, MacLeod D, Lorino H, Farre R, Hantos Z, Desager K, Marchal F et al (2003) The forced oscillation technique in clinical practice: methodology, recommendations and future developments. Eur Respir J 22(6):1026–1041
22. Pham TT, Thamrin C, Robinson PD, McEwan A, Leong PH (2016) Respiratory artefact removal in forced oscillation measurements: a machine learning approach. IEEE Trans Biomed Eng 64(7):1–9
23. Pham TT, Leong PH, Robinson PD, Gutzler T, Jee AS, King GG, Thamrin C (2017) Automated quality control of forced oscillation measurements: respiratory artifact detection with advanced feature extraction. J Appl Physiol 123(4):781–789
24. Que CL, Kenyon C, Olivenstein R, Macklem PT, Maksym GN (2001) Homeokinesis and short-term variability of human airway caliber. J Appl Physiol 91(3):1131–1141
25. Rijsbergen CJV (1979) Information retrieval, 2nd edn. Butterworth-Heinemann, Newton, USA
26. Robinson PD, Turner M, Brown NJ, Salome C, Berend N, Marks GB, King GG (2011) Proce-dures to improve the repeatability of forced oscillation measurements in school-aged children. Respir Physiol Neurobiol 177(2):199–206
27. Schölkopf B, Platt JC, Shawe-Taylor J, Smola AJ, Williamson RC (2001) Estimating the support of a high-dimensional distribution. Neural Comput 13(7):1443–1471
28. Schweitzer C, Chone C, Marchal F (2003) Influence of data filtering on reliability of respiratory impedance and derived parameters in children. Pediatr Pulmonol 36(6):502–508
29. Shannon C (1948) A mathematical theory of communication. Bell Syst Tech J 27(3):379–423
30. Smith H, Reinhold P, Goldman M (2005) Forced oscillation technique and impulse oscillometry. Eur Respir Monogr 31:72
31. Székely GJ, Rizzo ML, Bakirov NK et al (2007) Measuring and testing dependence by corre-lation of distances. Ann Stat 35(6):2769–2794
32. Thorpe CW, Salome CM, Berend N, King GG (2004) Modeling airway resistance dynamics after tidal and deep inspirations. J Appl Physiol 97(5):1643–1653

33. Timmins SC, Diba C, Thamrin C, Berend N, Salome CM, King GG (2012) The feasibility of home monitoring of impedance with the forced oscillation technique in chronic obstructive pulmonary disease subjects. Physiol Meas 34(1):67–81
34. Timmins SC, Coatsworth N, Palnitkar G, Thamrin C, Farrow CE, Schoeffel RE, Berend N, Diba C, Salome CM, King GG (2013) Day-to-day variability of oscillatory impedance and spirometry in asthma and COPD. Respir Physiol Neurobiol 185(2):416–424
35. Vural V, Fung G, Krishnapuram B, Dy J, Rao B (2006) Batch classification with applications in computer aided diagnosis. In: Machine learning: ECML 2006. Springer, pp 449–460
36. Wilcoxon F (1945) Individual comparisons by ranking methods. Biom Bull 80–83

Chapter 6
Spike Sorting: Application to Motor Unit Action Potential Discrimination

6.1 Background on Electromyography Motor Unit Analysis

6.1.1 MUAPs

Motor unit activity analysis provides crucial information towards diagnosis and treatment of neuromuscular disorders. In intramuscular electromyography data, when recording small voluntary contractions with a needle electrode, the electrical signal obtained is often a sum of more than one motor unit (MU) from the surrounding area of the needle tip. Therefore, a motor unit action potential (MUAP) consists of several muscle fiber action potentials (MFAPs) within the anatomical MU.

Single MU activity is of research interest because changes of MUAP morphology, MU activation, and MU recruitment yield valuable information. Neuropathic conditions occur with decreased recruitment whereas myopathic conditions happen with MUAP morphology changes. As an example, a MUAP examination can confirm myopathic conditions and identify the differential to find an appropriate biopsy site [22]. On the other hand, most neurology laboratories utilize experts who spend hours classifying action potentials ("spikes") using commercial software tools (e.g., Spike2 [9], Cerebus [6]) after each recording. Hence, an unsupervised method is highly desirable.

6.1.2 Spike Sorting

A practical spike discrimination procedure involves three basic phases: spike detection, feature extraction, and spike clustering. Spike detection often involves aligning spikes to a common temporal point. The feature extraction phase provides principal information that highlights differences among spikes. A dimensionality reduction

© Springer Nature Switzerland AG 2019
T. T. Pham, *Applying Machine Learning for Automated Classification of Biomedical Data in Subject-Independent Settings*, Springer Theses, https://doi.org/10.1007/978-3-319-98675-3_6

step is executed to select only the few best coefficients. In the final phase, spikes are assigned into different MU classes.

Spike classification processes include two main steps: extracting spike features, and then classifying spikes using these spike features. Common spike feature extraction algorithms are based on principal component analysis (PCA) [14] used [1, 13], the discrete wavelet transform (DWT) [19] applied [16, 25], independent component analysis (ICA), [28] found [21, 27], or discrete derivatives [18]. Other existing algorithms use waveform derivatives [29], the integral transform [30], inter-spike intervals [8], or Laplacian eigenmaps [5].

6.2   Data Collection

6.2.1   Physiologically Based Synthetic Data

We used the nEMG simulation algorithm by Hamilton-Wright and Stashuk [15] for development in this work. This was shown to produce nEMG data consistent with those acquired from real muscle (the developed muscle). We run the simulator on a Microsoft Windows personal computer for a concentric electrode during a 10% maximal voluntary contraction (MVC). Figure 6.1 illustrates a synthetic epoch of 100 ms.

6.2.2   Recorded Data

We also collected a real data set recording from a healthy young male at the Fuglevand Laboratory [10] using a rack-mounted electro-physiological recording system CED

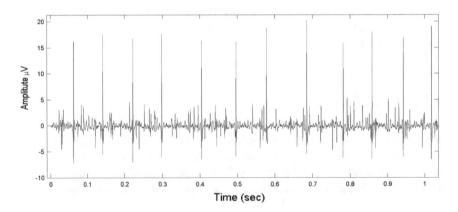

Fig. 6.1 Example of a 100-ms epoch of the simulated nEMG

[9]. Data were sampled at 55.5 kHz. The experiment settings for force used to create nEMG data was: time interval of 0.1 ms for force, scale of 0.0023, unit of "N". The electrode type was the concentric needle electrode. A neurologist manually provided labels of MUAP appearances together with its associated MU. Though most of the manual labeling procedure was aided by a commercial software tool (Spike2 [9]), the human operator is still needed for final template matching and adjusting. We call this labels "reference" during our evaluation.

6.3 Feature Pool

We extracted twelve groups of features in both time and frequency domains (Table 6.1). Existing features include amplitude range information of EMG data, DWT, top ten selected by KS tests [20] of DWT, top 10% selected by ICA or PCA. New feature candidates are singular value decomposition (SVD) of spectral analysis and spectrograms of raw amplitude data or DWT transformed data.

Several methods used for *new* feature extraction (i.e., have not been proposed for nEMG spike sorting) are described as follows. Discrete wavelet analysis that represents signals in both frequency and time is a very useful tool in the neuroscience field [26]. Transient differences in high frequency features (sharp edges and steep leading or trailing slopes) and/or in low frequency features (duration of the repolarization phase) can present the morphology of spikes. In this work, MUAPs are first decomposed into wavelet coefficients using the DWT method [19]. These coefficients represent differences among spikes based on the quantification of energy found in specific frequency bands at specific time locations (details in Appendix).

Table 6.1 List of candidates in the EMG feature pool. New*: Features have not been previously proposed

Group ID	Domain	Description	New?*	Feature ID
1	Time	Maximum amplitude of EMG	No	1
2	Time	Minimum amplitude of EMG	No	2
3	Time	Range amplitude of EMG	No	3
4	Frequency	DWT level d3	No	4–128
5	Frequency	DWT level d4	No	129–253
6	Frequency	DWT level a3	No	254–378
7	Frequency	SVD of spectral analysis	Yes	379–386
8	Time	ICA (10%)	No	387–398
9	Time	PCA (10%)	No	399–410
10	Frequency	KS test of DWT (top ten coefficients)	No	411–420
11	Frequency	Spectrograms of raw amplitude	Yes	421–1065
12	Frequency	Spectrograms of DWT	Yes	1066–1710

We implemented a 4-level decomposition and *Haar* window using built-in functions of MATLAB (The MathWorks Inc., Natick, MA, 2000).

Due to the multi-modal distribution of coefficients [25], we rank these candidates by scores calculated by deviation from normality, using a modification of Kolmogorov–Smirnov (KS) test [20]. Let X be a data set, the score is $max(|F(x)G(x)|)$ where $F(x)$ is the cumulative distribution function of X and $G(x)$ is a Gaussian cumulative distribution function with the same mean and variance. To minimize the effect of overlapping spikes, for each coefficient, only values within three standard deviations (both directions) are considered [25]. To better compare with existing most relevant methods, the ten largest score candidates are selected as previously suggested to separate spikes.

These selected coefficients are transformed to a series of spectral snapshots (spectrograms) using the short Fourier transform (STFT [12]). Specifically, let $v \in \mathbb{R}^{1 \times 10}$ be the wavelet feature of a spike. A Hamming window is used with STFT to transform v into an image of spectrogram I (e.g., 5×129). Hence, *distance* between spikes used in the next sorting process are the correlation coefficients between these images.

6.4   Automated Spike Sorter

6.4.1   *Preprocessing*

Intramuscular data is corrupted by spike-like correlated noise. Thus, we need to make data points statistically independent ("pre-whitening"). A practical approach employs a linear prediction filter [17] to whiten the input signal itself before we extract any MUAP. In this work, we use a third-order forward linear predictor (FIR filter) that predicts the current value of the real-valued original data based on past three samples [17]. Using timing labels from the reference, we extract the spike set together with labels of MU classes. All spikes are extracted with the same window size of 8 ms.

To focus on sorting evaluation, overlapping spikes (i.e., have more than two MU in the same window) relate more to spike detection than sorting algorithms. Thus, we removed overlapping spikes with small delay by detecting multiple peaks within a spike window. For overlaps without delay (i.e., they may look like the firing of a new neuron) we consider these spikes a separate class.

Because the firing behaviour of an individual MU relates to its recruitment threshold [2, 7], the size of a valid cluster corresponding to a MU should exceed a parameter. According to the recruitment threshold assignment derived from the work of Fuglevand [11] and popular settings found in the literature, we set this parameter to 40. All clusters with size smaller than 40 were merged into a group, called *catch-all* class. In the previous works of sorting performance evaluation, this class is sometimes set apart. We assume that these small clusters may associate with overlapping spikes without delay. Thus, we evaluate two cases of detection performance: *include catch-all* and *exclude catch-all*. To assign the label for cluster (or individual spike if that is the catch-all cluster), we measure the correlation between the mean waveform of the cluster and the one of the reference group. A label is chosen if the match has the highest correlation score.

6.4.2 x-Class Sorter

After feature extraction steps, based on MUAP morphology, the correlation between spikes is used as the similarity measure for an number x-class sorting application [23, 24] where x is unknown. Rather than using the Euclidean distance metric, to account for electrode drift and subject-independent setting requirements, the correlation metric that ranges $-1 \rightarrow 1$ is used.

Let I_X and I_Y be two feature vectors of MUAP X and MUAP Y, respectively. $r_{X,Y}$ is the correlation between two feature vectors of X and Y (Eq. (3.7)). The class assignment variable of X is defined by the correlation based sorting scheme (Sect. 3.2.3). The sorter starts with a single class contains all spikes having high correlation $r_{X,Y}$ with the initial spike given a desired threshold level (e.g., 0.9). Then the sorter stops when the unsorted pool of remaining spikes is empty.

6.5 Reference Works

The objective reference clustering results are available for the synthetic data as the simulator is controlled during data generation. However, this is usually not available for the recorded data. Ideally the reference could be derived from simultaneous intracellular recording, but availability of such data is limited. The most common practice in physiology laboratories involves using commercial software (specifically *Spike2* in our work) and manual checking by a human operator. This was the approach used to obtain the reference for our real recorded data.

We also compare our proposed method with a relevant work using the DWT extraction and super paramagnetic clustering (SPC) [3]. We applied the default settings for the SPC method as recommended by Blatt et al. [4]. Specifically there were $q = 20$ states, $K = 11$ nearest neighbours, and $N = 500$ iterations for clustering. The range of temperature was from 0 to 0.201 in steps of 0.01. The implementation was provided by the authors [3] (MATLAB packages, The MathWorks Inc., Natick, MA, 2000).

6.6 Performance Metrics

Performance metrics for a multi-class classification task are derived from the confusion matrix. Let M be the confusion matrix of sorting outcome (Eq. (2.3) in Chap. 2). The successful predicted events (*True*) for a class are on the diagonal of M. All other members of M are incorrectly predicted events (*False*).

Let $M_{i,j}$ denote the number of test outcomes (i.e., *ground truth* labels, G_i) of class i, that were predicted as class j, P_i. The number of successful predicted events (*True*) for class i, denoted T_{ii}, is the diagonal of M. All other members of M are incorrectly predicted events (*False*), denoted F_{ij} where $i \neq j$.

$$M = \begin{matrix} P_1 & \ldots & P_i & \ldots & P_C & \\ \begin{pmatrix} T_{11} & \ldots & F_{1i} & \ldots & F_{1C} \\ \vdots & \ddots & \vdots & \ldots & \vdots \\ F_{i1} & \ldots & T_{ii} & \ldots & F_{iC} \\ \vdots & \ldots & \vdots & \ddots & \vdots \\ F_{C1} & \ldots & F_{Ci} & \ldots & T_{CC} \end{pmatrix} & \begin{matrix} G_1 \\ \vdots \\ G_i \\ \vdots \\ G_C \end{matrix} \end{matrix} \tag{6.1}$$

The sensitivity and positive predictive value (PPV) of class i, Sen_i and PPV_i, are defined as follows.

$$\mathrm{Sen}_i = \frac{T_{ii}}{T_{ii} + \sum_{j \neq i} F_{ij}} \tag{6.2}$$

$$\mathrm{PPV}_i = \frac{T_{ii}}{T_{ii} + \sum_{j \neq i} F_{ji}} \tag{6.3}$$

6.7 Results

6.7.1 Selected Features

Figure 6.2 illustrates the ranking scores by saliency criteria for each feature candidate over the entire exploratory pool (sorted from high to low scores); a higher saliency score indicated the higher ranking order. The top highest-score candidates were investigated using histograms by feature Group ID (Table 6.1) as in Fig. 6.3.

As can be seen, scores dropped quickly outside of the top 25% candidates by DIS criterion and only after 80% candidates by the MI score. We noticed that, by MI criterion, except for the single feature of Group 1, all other members of the top 25% belong to Group 12. Meanwhile, by DIS criterion, though the top 25% includes several groups, Group 12 still dominates the high score area. Hence, we proposed to use the feature set Group 12 for the next evaluation in terms of sorting performance.

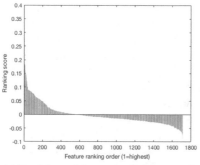
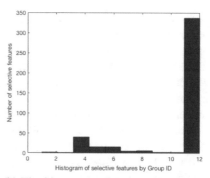

(a) Ranking of the entire feature pool.   (b) The histogram of the top 25 percent
                                                                  candidates.

Fig. 6.2 Feature ranking results by DIS criterion. **a** Ranking scores for the entire feature pool. Vertical axes: scores calculated by saliency criteria; Horizontal axes: ranking order (highest = 1, lowest = 1710). **b** The histogram of the top 25% highest-score candidates by feature groups (Table 6.1)

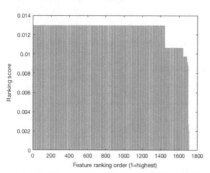
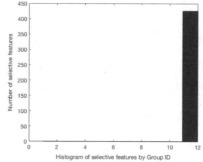

(a) Ranking of the entire feature pool.   (b) The histogram of the top 25 percent
                                                                  candidates.

Fig. 6.3 Feature ranking results by MI criterion. **a** Ranking scores for the entire feature pool. Vertical axes: scores calculated by saliency criteria; Horizontal axes: ranking order (highest = 1, lowest = 1710). **b** The histogram of the top 25% highest-score candidates by feature groups (Table 6.1)

6.7.2 Sorting Performance

6.7.2.1 Synthetic Data

After preprocessing, spike sets were prepared for the sorting stage as in Table 6.2. Note that, though a confusion matrix may give much more information on misclassification, in the context of this thesis, in all applications we presented there is no known ground truth, the best practice has been subjective manual labels. Thus, we used visualization tools to compare our observations. Figure 6.4 depicts five reference classes (Fig. 6.4a) and two clustering results using the SPC sorter (Fig. 6.4b) and our proposed sorter (Fig. 6.4c). The MU1 class has much larger amplitude range than other four classes in the reference set. Classes MU2-5 have only slight difference in the waveforms. In the SPC clustering result, a class may include more than one cluster (e.g., class MU2 and MU4 in Fig. 6.4b). Also, the MU5 class may be included in clusters of other classes. Our sorter produced five clusters that match with five reference classes though it does have a catch-all group similar to the SPC method.

The proportion of the catch-all group is reported in Fig. 6.5a. After assigning labels, the histograms were compared with the reference histogram (Fig. 6.5b). In terms of the confusion matrix, the general classification accuracy and class-wise sensitivities as well as predictivities are reported in Table 6.3.

6.7.2.2   Recorded Data

Table 6.4 depicts the distribution of spikes in large clusters corresponding to the reference classes from the recorded dataset. Both automatic clustering methods had

Table 6.2 Class proportions of spike set inputs are in order of the MU names in the labels

	Synthetic data	Recorded data
No. data points (sampling rate)	937500 (31 kHz)	7500000 (55 kHz)
Number of spikes	1230	1220
Number of classes	5	3
Class proportion	336:269:226:207:192	440:483:535

Table 6.3 Synthetic data MUAP sorting comparisons between automatic methods and the reference. Accuracy measures (in %) use simulation settings as reference. True/False are MU matching or not with the reference labels

Metrics	Class	Include catch-all		Exclude catch-all	
		SPC-based	Our method	SPC-based	Our method
Sensitivity	MU1	39.3	92.5	100.0	100.0
	MU2	85.1	67.6	100.0	98.4
	MU3	74.7	75.2	98.8	98.3
	MU4	71.9	69.5	99.3	88.9
	MU5	19.3	88.0	0	72.9
PPV	MU1	100.0	100.0	100.0	100.0
	MU2	81.2	98.9	98.2	98.9
	MU3	54.5	80.2	54.5	80.2
	MU4	98.6	97.9	54.6	97.9
	MU5	10.4	44.9	0	100.0
Average accuracy		58.2	79.3	81.9	94.8

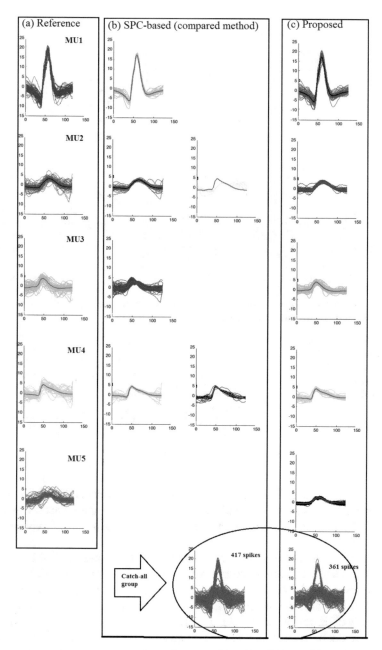

Fig. 6.4 Clustering results using synthetic data. **a** There are five classes from the reference labels (MU1 to MU5). **b** SPC sorter (compared method) may include more than one cluster for the same class. **c** Our proposed method using correlation based clustering. Axes x, y are time index and amplitude of the spikes (μV), respectively. Different colours are different clusters made by sorters, not colour coded for the class

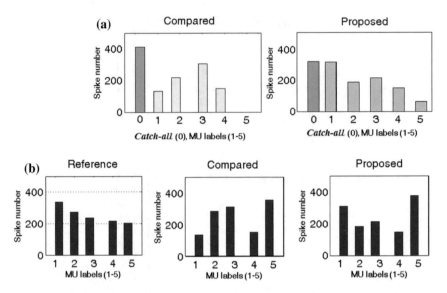

Fig. 6.5 Agreement in histograms of automatic methods against the reference for synthetic data. **a**: excluded *catch-all* **b**: included *catch-all*

Table 6.4 Distribution of spike count in large clusters corresponding to the reference classes from the recorded dataset

	Our method	The SPC-based	The reference
MU1 MUAPs	314	368	383
MU2 MUAPs	283	335	408
MU3 MUAPs	378	284	429
Remaining MUAPs	245	233	0

about 19% *catch-all* spikes. There were three reference classes. While the amplitude range of spikes in MU1 and MU2 are $\pm 0.5\,\mu V$, MU3 ranges are much larger ($\pm 1\,\mu V$). Sorting performance for each MU and the general accuracy were depicted in Table 6.5.

In contrast with our superior results against the SPC when applied to the synthetic data, results of both automatic techniques were comparable with recorded data. However, the size of dataset as well as a small number of active MUs recorded may explain for this. We may also need an inter-rater measurement to alleviate the subjectivity of the reference in evaluation. These should be addressed in future work for the method. In general, all performance measurements we achieved in this study are among the most accurate outcomes in spike sorting evaluation works (Fig. 6.5).

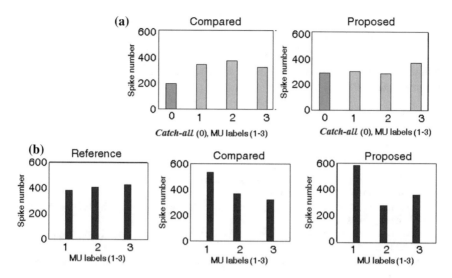

Fig. 6.6 Agreement in histograms of automatic methods against the reference for recorded data. **a** Exclude *catch-all* **b** Include *catch-all*

Table 6.5 Comparison of sorting performance using recorded data between automatic methods. Accuracy measures (in %) use manual labels as reference. True/False are MU matching or not with the reference labels

Metrics	Class	Include catch-all		Not include catch-all	
		The SPC-based	Our method	The SPC-based	Our method
Sensitivity	MU1	99.7	99.7	99.7	99.6
	MU2	89.9	68.6	100.0	99.6
	MU3	74.6	83.9	100.0	100.0
PPV	MU1	71.8	65.8	100.0	99.6
	MU2	99.7	99.6	99.7	99.6
	MU3	100.0	100.0	100.0	100.0
General accuracy		87.6	83.7	99.9	99.8

6.8 Summary

In this chapter, a classification application where the number of classes is not known is reported. A similar feature ranking can still be used to better extract features. The extention of using anomaly scores in the classifier was also illustrated. Synthetic and real recorded datasets of motor unit action potentials were used to evaluate the performance. Comparing with the manual reference, our MUAP classification method is comparable (regarding to the number of MUs found and histograms of MUs). Moreover, in the SPC method, the default settings assumed a maximum number of clusters. If the real recording conditions provoke more classes than that parameter, a technical specialist may need to redefine the parameter. Furthermore, the *temperature* terminology used in the SPC for reviewing outcome is less intiuitive than the correlation as in our method. The correlation values range $-1 \rightarrow 1$ while the measure of *temperature* is difficult to tune.

References

1. Abeles M, Goldstein MH (1977) Multispike train analysis. Proc IEEE 65(5):762–773. https://doi.org/10.1109/PROC.1977.10559
2. Basmajian JV, De Luca CJ (1985) Muscles alive: their functions revealed by electromyography. Williams & Wilkins
3. Blatt M, Wiseman S, Domany E (1996) Superparamagnetic clustering of data. Phys Rev Lett 76:3251–3254
4. Blatt M, Wiseman S, Domany E (1997) Data clustering using a model granular magnet. Neural Comput 9:1805–1842
5. Chah E, Hok V, Della-Chiesa A, Miller JH, O'Mara SM, Reilly RB (2011) Automated spike sorting algorithm based on Laplacian eigenmaps and k-means clustering. J Neural Eng 8:016,006 (9 pp)
6. Cyberkineticsinc (-) Cerebus. www.cyberkineticsinc.com
7. De Luca CJ (1979) Physiology and mathematics of myoelectric signals. IEEE Trans Biomed Eng 6:313–325
8. Delescluse M, Pouzat C (2006) Efficient spike-sorting of multi-state neurons using inter-spike intervals information. J Neurosci Methods 150(1):16–29
9. Design CE (-) Spike2. www.ced.co.uk
10. Fuglevand A (-) Fuglevand laboratory of motor control neurophysiology. Department of Physiology, University of Arizona, USA
11. Fuglevand AJ, Winter DA, Patla AE (1993) Models of recruitment and rate coding organization in motor-unit pools. J Neurophysiol 70(6):2470–2488
12. Gabor D (1946) Theory of communication. Part 1: the analysis of information. J Inst Electr Eng. Part III Radio Commun Eng, 429–441
13. Glaser EM (1971) On-line separation of interleaved neuronal pulse sequences. Data Acquis Process Biol Med 1:77–136
14. Glaser EM, Marks WB (1968) Separation of neuronal activity by waveform analysis. Adv Biomed Eng 5:137–156
15. Hamilton-Wright A, Stashuk DW (2005) Physiologically based simulation of clinical EMG signals. IEEE Trans Biomed Eng 52(2):171–183
16. Hulata E, Segev R, Ben-Jacob E (2002) A method for spike sorting and detection based on wavelet packets and Shannon's mutual information. J Neurosci Methods 117(1):1–12
17. Jackson LB, et al (1989) Digital filters and signal processing, vol 3. Springer
18. Karkare V, Gibson S, Markovic D (2009) A 130 μW, 64-channel spike-sorting DSP chip. In: IEEE Asian solid-state circuits conference 2009, A-SSCC 2009, pp 289–292
19. Letelier JC, Weber PP (2000) Spike sorting based on discrete wavelet transform coefficients. J Neurosci Methods 101(2):93–106
20. Lilliefors HW (1967) On the Kolmogorov–Smirnov test for normality with mean and variance unknown. J Am Stat Assoc 62(318):399–402
21. Mamlouk AM, Sharp H, Menne KM, Hofmann UG, Martinetz T (2005) Unsupervised spike sorting with ICA and its evaluation using GENESIS simulations. Neurocomputing 65–66:275–282. Computational Neuroscience: Trends in Research 2005

22. Paganoni S, Amato A (2013) Electrodiagnostic evaluation of myopathies. Phys Med Rehabil Clin North Am 24(1):193–207
23. Pham TT, Higgins CM (2014) A visual motion detecting module for dragonfly-controlled robots. In: 36th annual international conference of the IEEE engineering in medicine and biology society (EMBC) 2014. IEEE, pp 1666–1669
24. Pham TT, Fuglevand AJ, McEwan AL, Leong PH (2014) Unsupervised discrimination of motor unit action potentials using spectrograms. In: 36th annual international conference of the IEEE engineering in medicine and biology society (EMBC). IEEE, pp 1–4
25. Quiroga R, Nadasdy Z, Ben-Shaul Y (2004) Unsupervised spike detection and sorting with wavelets and superparamagnetic clustering. Neural Comput 16:1661–1687
26. Samar VJ (1999) Wavelet analysis of neuroelectric waveforms. Brain Lang 66:1–6
27. Snellings A, Anderson DJ, Aldridge JW (2006) Improved signal and reduced noise in neural recordings from close-spaced electrode arrays using independent component analysis as a preprocessor. J Neurosci Methods 150(2):254–264
28. Takahashi S, Anzai Y, Sakurai Y (2003) A new approach to spike sorting for multi-neuronal activities recorded with a tetrode-how ICA can be practical. Neurosci Res 46(3):265–272
29. Yang Z, Zhao Q, Liu W (2009) Improving spike separation using waveform derivatives. J Neural Eng 6:046,006–046,018
30. Zviagintsev A, Perelman Y, Ginosar R (2005) Low-power architectures for spike sorting. In: Proceedings of the 2nd international IEEE EMBS conference on neural engineering, pp 162–165

Chapter 7
Conclusion

This chapter summarises the proposed feature engineering method for classification applications (Chap. 3) and the main observations in several experiments presented in Chaps. 4, 5 and 6. Discussions also include limitations and future works for each scenario. In general, the contribution of a systematic application of feature engineering to accuracy performance is shown in all three cases of real-life biomedical data classification.

7.1 Proposed Algorithms

This thesis proposed classification schemes for unsupervised and subject-independent settings in biomedical data processing applications, especially for automated deployments in out-of-the-lab environments. This not only helps eliminate the subjectivity associated with human involvement, but it also reduces labour costs.

Existing automated efforts have been predominantly designed for subject dependence and only yielded modestly accurate results for subject-independent settings. In this thesis, three examples (human body movement assessment (Chap. 4), respiratory artefact removal (Chap. 5), and spike sorting for electrophysiological data (Chap. 6) demonstrated that the classification performance of unsupervised and subject-independent automated sorters for biomedical data can be improved by exploiting data-driven and domain-knowledge-driven strategies that help find better features and more efficient sorters.

© Springer Nature Switzerland AG 2019
T. T. Pham, *Applying Machine Learning for Automated Classification of Biomedical Data in Subject-Independent Settings*, Springer Theses, https://doi.org/10.1007/978-3-319-98675-3_7

7.1.1   Feature Engineering

This thesis improved data mining in subject-independent settings by using supervised techniques to find better features (i.e., more discriminative and higher correlated with the desired output). A voting-based technique has been proposed to analyze ranking scores by several saliency criteria including mutual information, Euclidean distance based discrimination, and variance ratio based clusterability. This hybrid selection scheme is a data-driven approach and can compare a comprehensive set of candidates including existing features and novel variants. Given a large set of exploratory feature candidates, the most selective features learnt from this process are most applicable to the unsupervised and subject-independent applications.

The feature selection technique based on voting has been first reported for respiratory artefact removal in FOT measurements [2–4], FoG detection [5–7], and nEMG spike sorting [1]. The voting selection considers not only mutual information criterion but also clusterability. Novel efficient features were discovered thanks to the fact that they are more relevant and discriminative than existing ones commonly used in the FoG, FOT, and nEMG literature [1, 3, 5].

7.1.2   Classifiers

In each application, better models have been suggested through this domain-knowledge-driven approach (e.g., issues associated with dependency in Chap. 5 and/or other related domain knowledge in Chaps. 4 and 6). Specifically, in Chaps. 4 and 5, the proposed feature learning resulted in anomaly detectors which, to the best of our knowledge, achieve the best reported performance for unsupervised subject-independent settings for FOT data regardless of participants' age [4] and FoG data [6, 7]. In Chap. 6, an efficient unsupervised spike sorter is introduced when the class number is not known for subject-independent settings [1].

7.2   Experiment Results

7.2.1   Point Anomaly Detection Application

As freezing of gait instance of patients with Parkinson's disease can be detected as point anomalies, relevant and discriminative features can make simple thresholding filters work as detectors.

According to the feature ranking in Chap. 4, apart from the existing features (e.g., the freezing index extracted from ankle sensor at vertical axis), the new feature with multiple channels, FI_{MC}, is one of the top features in saliency, clusterability, and robustness. Only seven out of 244 exploratory candidates met requirements of our three-round selection procedure.

The proposed anomaly score based detector, ASD, is a simple thresholding method but using dynamic threshold values that make ASD suitable for subject-independent requirements. In Chap. 4, the ASD method significantly outperformed existing works with a small window and/or low tolerance. For example, for *ASD ankle* (y-axis), the mean ($\pm$SD) of sensitivity, specificity are 94% ($\pm$23%) and 84% ($\pm$36%) while the recent work [8] only achieved 75% and 76%, respectively.

These findings form a further step towards subject-independent out-of-lab FoG detectors. In future work, a combination of top ranking features should be further evaluated. A more elaborate technique for the ASD threshold settings is also worthy of further study.

7.2.2 Collective Anomaly Detection Application

In Chap. 5, another anomaly detection scenario is demonstrated using data from lung function tests. Breath cycles that include any respiratory artefact data points are required to be removed (to appropriately assess the lung function). A complete-breath removal approach is applied to ensure the balance of a cycle. Hence, the removal can be done by a collective anomaly detection technique. This thesis utilized data sets of forced oscillation technique (FOT) recorded from adults and eight- to eleven-year-old children in Queensland and New South Wales, Australia.

Based on observations from the proposed feature ranking steps, a new set of *landmark* features were proposed. These were extracted from boundary points of two-dimensional resistance-against-flow curves. This feature group is highly ranked by supervised learning techniques using saliency scores (DIS, MI, variance ratio). The MI score measures the correlation (mutual information) between one feature candidate of a breath and its label of abnormality. Meanwhile, DIS and variance ratio scores depict the clusterability of a feature candidate.

Although selecting the ten highest score candidates is common practice in the feature learning literature, an investigation of the stability of these feature preferences should be undertaken. Nevertheless, our results are consistent with more than one well-known feature selection algorithm with four separate data sets. As reported in Sect. 5.8.1, scores that come after the top ten were significantly lower than the top ten group. Thus, $k = 10$ satisfied our requirements. In practice, one may choose the entire landmark group and the resulting detector will perform comparably to the approach of this thesis. This is because the majority of the top ten are actually landmark features and the performance curves varied negligibly among selection algorithms.

While we demonstrated a reasonable degree of independence between the accuracy of our detector and levels of obstruction (details in Sect. 5.8.2.2), further work is required to determine if the detector can be applicable to recordings from severely obstructed patients or those experiencing an exacerbation. Also of note is that in our datasets of healthy and asthmatic subjects, *Rrs* features ranked consistently high, whereas the features associated with *Xrs* did not rank highly for inclusion in the detector. This may be different in other diseases, and remains to be tested.

Finally, the analysis was limited to a single frequency closest to what is usually reported in the literature (5 Hz). However, the detector could also be applied to multi-frequency systems which are commonly used, using a similar set of features for each component frequency.

In terms of detection performance, we used several metrics (e.g., ROC, throughput, and variability (Sect. 5.3)) to determine threshold parameters. During development, the performance curves (i.e., F1-scores, ROC, and the variability) against the parameter n_{IQR} showed that the top ten features outperformed the case of no feature selection. The three saliency scores yield nearly similar performance curves. The proposed artefact detector, *IIQR-MI* achieved promising results in subject-independent settings, *regardless of age*. In out-of-sample tests, our detector performed similar to the *gold standard*, as assessed through paired t-tests (two-tailed) for variability.

Our findings are an important first step towards objective and automated quality control of FOT measurements, as FOT moves beyond its long-standing role in the respiratory research realm, becomes more available in commercial systems and is increasingly adopted in clinical and home telemonitoring settings.

7.2.3 Spike Sorting Application

In Chap. 6, we applied the proposed feature ranking scheme to a different classification scenario where the number of classes is unknown. *Spike sorting* is a typical example for this case where the main task is to discriminate motor unit action potentials using nEMG data. Both types of data sources were used: synthetic and real recorded nEMG recordings from human subjects.

From the feature ranking observation, a novel candidate set has been suggested as it was higher correlated to the motor unit reference and was more separable than existing features. Then the Chapter introduced a correlation based clustering technique. Compared with the reference produced by human experts, the proposed method obtained a comparable result. The number of classes was found to be equivalent. MUAP morphology was identical in each pair of corresponding MU class, and the histograms of MUs by the proposed method were also similar to the reference ones.

7.3 Summary

Technical background and details of proposed algorithms for feature relevance selection as well as classifiers were discussed in Chaps. 2 and 3. Then biomedical background and their literature review for the three application scenarios were provided in the remaining chapters of this thesis. Chapter 4 illustrated point anomaly detection in human body movement assessment using accelerometer data for FoG in patients with advanced Parkinson's disease. Chapter 5 reported a collective anomaly detection case study using lung function test data. Chapter 6 presents outcomes of multi-class classification in spike sorting for motor unit action potential in nEMG data.

Summary of Findings

1. Current objective methods for FoG detection used various global parameters and/or different channels. This suggests a high variability in actual thresholds over time and subjects.
2. The averages of FOT measurements, which are the main outcomes in clinical and research applications, are affected significantly by the artefacts. Apart from the natural dependency of breath samples, the normality assumption of data within a recording is invalid by current hypothesis tests. Hence, beside choosing better features, more general statistical parameters with quartiles should be applied rather than existing methods with the normality assumption.
3. Though single MU activities provide the most informative part in diagnosis and treatment of neuromuscular disorders, nEMG data often provide more than one MU activities. Thus MUAP discrimination is a crucial task. Note that the number of classes in this classification task is unknown. Hence a well-suited metric to sort is the correlation between MUAP waveforms.
4. The de facto standard or ground-truth practice for these three cases has been the manual sorting that is laborious and subjective. Unsupervised methods using simple statistical thresholds have only yielded modest performances. Supervised learning models have been mainly reported with excellent results for subject-based rather than subject-independent settings.
5. New features found by the feature engineering could help deploy a low computational cost classifier and thus make it more generalized with respect to subject variations.

The three real-life applications demonstrated in this thesis illustrate that systematic feature engineering could help replace standard manual classification with automated classifiers that are unsupervised, subject-independent, and of low computational cost.

References

1. Pham TT, Fuglevand AJ, McEwan AL, Leong PH (2014) Unsupervised discrimination of motor unit action potentials using spectrograms. In: 36th Annual international conference of the IEEE Engineering in medicine and biology society (EMBC) 2014. IEEE, pp 1–4
2. Pham TT, Nguyen DN, Dutkiewicz E, McEwan AL, Thamrin C, Robinson PD, Leong PH (2016a) Feature engineering and supervised learning classifiers for respiratory artefact removal in lung function tests. In: Global communications conference (GLOBECOM). IEEE, pp 1–6
3. Pham TT, Thamrin C, Robinson PD, McEwan A, Leong PH (2016b) Respiratory artefact removal in forced oscillation measurements: a machine learning approach. IEEE Trans Biomed Eng 64(7):1–9

4. Pham TT, Leong PH, Robinson PD, Gutzler T, Jee AS, King GG, Thamrin C (2017a) Automated quality control of forced oscillation measurements: respiratory artifact detection with advanced feature extraction. J Appl Physiol 123(4):781–789
5. Pham TT, Moore ST, Lewis SJG, Nguyen DN, Dutkiewicz E, Fuglevand AJ, McEwan AL, Leong PH (2017b) Freezing of gait detection in Parkinson's disease: a subject-independent detector using anomaly scores. IEEE Trans Biomed Eng 64(11):2719–2728
6. Pham TT, Nguyen DN, Dutkiewicz E, McEwan AL, Leong PH (2017c) An anomaly detection technique in wearable wireless monitoring systems for studies of gait freezing in Parkinson's disease. In: 2017 International conference on information networking (ICOIN). IEEE, pp 41–45
7. Pham TT, Nguyen DN, Dutkiewicz E, McEwan AL, Leong PH (2017d) Wearable healthcare systems: A single channel accelerometer based anomaly detector for studies of gait freezing in Parkinson's disease. In: IEEE international conference on communications (ICC). IEEE, pp 1–5
8. Zach H, Janssen AM, Snijders AH, Delval A, Ferraye MU, Auff E, Weerdesteyn V, Bloem BR, Nonnekes J (2015) Identifying freezing of gait in Parkinson's disease during freezing provoking tasks using waist-mounted accelerometry. Parkinsonis Relat Disord 21(11):1362–1366

Appendix A
Wavelet Decomposition and Spectral Coherence

A.1 Wavelet Decomposition

Wavelet decomposition coefficients (DWT) [1] and spectral coherence [2] was calculated as below. Let $s(t)$ be a curve which can be presented by coefficients $C(a, b)$ (A.1).

$$C(a, b) = \frac{1}{\sqrt{a}} \int_{-\infty}^{+\infty} s(t)\psi_{a,b}(t)dt \qquad (A.1)$$

where $\psi_{a,b}(t) = \psi\left(\frac{t-b}{a}\right)$ is an expanded or contracted and shifted version of a unique wavelet function $\psi(t)$ a and b are the scale and the time localization, respectively.

A.2 Spectral Coherence

Let C_{XY} be the spectral coherence between signals X and Y. C_{XY} is defined by the Welch method [2] as in Eq. (A.2).

$$C_{XY}(\omega) = \frac{P_{XY}(\omega)}{\sqrt{P_{XX}(\omega).P_{YY}(\omega)}} \qquad (A.2)$$

where ω is frequency, $P_{XX}(\omega)$ is the power spectrum of signal x, $P_{YY}(\omega)$ is the power spectrum of signal y, and $P_{XY}(\omega)$ is the cross-power spectrum for signals x and y. When $P_{XX}(\omega) = 0$ or $P_{YY}(\omega) = 0$, then also $P_{XY}(\omega) = 0$ and we assume that $C_{XY}(\omega)$ is zero. To estimate power and cross spectra, let $\mathfrak{F}_x(\omega)$ and $\overline{\mathfrak{F}_x(\omega)}$, denote the Fourier transform and its conjugate of signal x, respectively, i.e. $\mathfrak{F}_x(\omega) = \int_{-\infty}^{+\infty} x(t).e^{-j\omega t}dt$. The power spectrum is then: $P_{XX}(\omega) = \mathfrak{F}_x(\omega).\overline{\mathfrak{F}_x(\omega)}$; $P_{YY}(\omega) = \mathfrak{F}_y(\omega).\overline{\mathfrak{F}_y(\omega)}$; and $P_{XY}(\omega) = \mathfrak{F}_x(\omega).\overline{\mathfrak{F}_y(\omega)}$.

© Springer Nature Switzerland AG 2019

T. T. Pham, *Applying Machine Learning for Automated Classification of Biomedical Data in Subject-Independent Settings*, Springer Theses, https://doi.org/10.1007/978-3-319-98675-3

A.3   STFT

: is the mathematical technique to produce spectrograms. Let $x[n]$ be an input vector to be transformed. $x[n]$ is broken up into frames (size m). Frames should overlap each other to avoid artefacts at the boundary. This transform can be expressed as

$$X(m, \omega) = \sum_{n=-\infty}^{\infty} x[n]h[n - m]e^{-j\omega n} \tag{A.3}$$

$$\text{spectrogram}\{x(n)\}(m, \omega) \equiv |X(m, \omega)|^2, \tag{A.4}$$

where $x[n]$ is an input of the transform, $h[n]$ is a window function with size m.

Appendix B
Table of Settings for Synthetic nEMG Data in Chapter 6

See Table B.1.

Table B.1 Settings for synthetic nEMG data

Number of motor units in muscle	200
Neuropathic MU loss fraction	0
Max adoption distance In um	150
Neuropathic MU enlargement fraction	1.5
Myopathic fibre affected fraction	0
Myopathic new involvement percentage in each cycle	5
Myopathic fibre gradually dying?	0
Dying and splitting depending on affection procedure?	1
Myopathic threshold of fibre death	25
Percentage of affected fibers dying	0
Myopathic fraction of fibres becoming hypertrophic	0.05
Factor of original area at which hypertrophic fibres split	2
Percentage of hypertrophic fibers splitting	0
Myopathic rate of atrophy	0.96
Myopathic rate of hypertrophy	1.04
Tip uptake distance	4500
Cannula uptake distance	4500
Radius of cannula shaft	250
Cannula length (in mm)	10
Needle X position (in mm)	0

(continued)

© Springer Nature Switzerland AG 2019

T. T. Pham, *Applying Machine Learning for Automated Classification of Biomedical Data in Subject-Independent Settings*, Springer Theses, https://doi.org/10.1007/978-3-319-98675-3

Table B.1 (continued)

Tip/Cannula reference setup	Tip versus Cannula
Needle Y position (in mm)	0
Needle Z position from NMJ in mm	15
Enable jitter?	True
Jitter (variance) in us	25
MFP threshold for jitter OR MU GST inclusion threshold (kV/s^2)	10
Minimum metric to seek needle to	0.25
Generate noise?	true
S/N ratio?	25
Recorded muscle name	Biceps brachii
Laterality	Right
Maximum recruitment threshold	50
Total time for EMG generation	29
Max (scaled) value in 16-bit output	4096
Internal interp. factor for jitter	30
Muscle fibre density	10
Area of 1 muscle fibre	0.0025
Min motor unit diameter	2
Max motor unit diameter	8
Ipi firing slope	0.8
Min firing rate	8
Max firing rate	42
Coeff of variance	0.25

About the Author

Thuy T. Pham earned her Ph.D. degree in Electrical and Information Engineering at The University of Sydney, NSW, Australia. Her passion for signal processing started at The University of Arizona, USA with her MSc in Electrical and Computer Engineering in 2010, where she built her background on Neuroscience, Robotics, Digital Signal and Image Processing. She was awarded the UA Meritorious Awards (2010–2012) and the Research Grant, Neuroscience Department, for her thesis on a real-time neural signal processing system for dragonflies towards hybrid bio-robots, at The Higgins Laboratory, Neuromorphic Vision and Robotic Systems, The University of Arizona. She received the Australian Prime Minister Award (Endeavour, 2013–2017) and Norman I. Price Scholarship for her Ph.D. study.

She has been held visiting scholar appointments and collaborated with world-leading medical and physiology groups: The Fuglevand Laboratory on Motor Control Neurophysiology, USA; Brain and Mind Research Institute, Sydney Medical School; Woolcock Institute of Medical Research, NSW; Garvan Institute of Medical Research, Darlinghurst NSW Australia.

References

1. Daubechies I, Bates BJ (1993) Ten lectures on wavelets. J Acoust Soc Am 93(3):1671–1671
2. Challis R, Kitney R (1990) Biomedical signal processing (part 3 of 4):the power spectrum and coherence function. Med Biol Eng Comput 28(6):509–524

Printed in the United States
By Bookmasters